Pocket Guide to

Public Speaking

Pocket Guide to Public Speaking

William Sanborn Pfeiffer
Southern Polytechnic State University

Upper Saddle River, New Jersey
Columbus, Ohio

Library of Congress Cataloging-in-Publication Data
Pfeiffer, William S.
Pocket guide to public speaking/William S. Pfeiffer.
p. cm.
Includes index.
ISBN 0-13-041544-8
1. Public speaking. I. Title.

PN4121 .P45 2002
808.5'41—dc21

2001021967

Editor in Chief: Stephen Helba
Executive Editor: Frank I. Mortimer, Jr.
Editorial Assistant: Barbara Rosenberg
Production Editor: Louise N. Sette
Production Supervision: Carlisle Publishers Services
Design Coordinator: Robin G. Chukes
Cover Designer: Thomas Borah
Cover photo: © SuperStock
Production Manager: Brian Fox
Marketing Manager: Jimmy Stephens

This book was set in Meridien by Carlisle Communications, Ltd. It was printed and bound by R. R. Donnelley & Sons Company. The cover was printed by Phoenix Color Corp.

Prentice-Hall International (UK) Limited, *London*
Prentice-Hall of Australia Pty. Limited, *Sydney*
Prentice-Hall Canada, Inc., *Toronto*
Prentice-Hall Hispanoamericana, S.A., *Mexico*
Prentice-Hall of India Private Limited, *New Delhi*
Prentice-Hall of Japan, Inc., *Tokyo*
Prentice-Hall Singapore Pte. Ltd.
Editora Prentice-Hall do Brasil, Ltda., *Rio de Janeiro*

10 9 8 7 6 5 4 3 2 1
ISBN: 0-13-041544-8

Preface

Good speeches should start by giving listeners three main pieces of information:

Purpose: What *purpose* does the speech have?

Importance: What *importance* does the speech have?

Plan: What *plan* will the speech follow?

It may not surprise you that this **PIP** formula applies not only to speeches but also to writing—from business proposals to handbooks like this one. So without delay, let's examine the purpose, importance, and plan for this book.

PURPOSE OF THE POCKET GUIDE TO PUBLIC SPEAKING

This little book aims to help three main groups improve their skills in public speaking:

1. Students in introductory speech classes that focus less on reading about speeches and more on preparing and delivering them
2. Students in other courses that require speeches and thus may use the book as a secondary text
3. Employees needing an on-the-job reference guide about speeches or a handbook for seminars on the subject

Not many people in any of these groups actually look forward to public speaking. Speeches force us outside our usual "comfort zone" of interpersonal, informal communication and into a "risky zone" of communication with larger groups in more formal situations. But because most of us *do* enjoy speaking informally, one key to good public speaking is to transfer skills we already display in less stressful contexts—such as conversations with friends or meetings with colleagues—to the formal contexts covered in this book. The *Pocket Guide to Public Speaking* will help you make this transition.

You've probably heard that every obstacle in life can be viewed conversely as an opportunity for learning. This book encourages you to see every speaking event as an opportunity to educate your audience. Public speaking also helps you grow personally and professionally. Just as you learn more about a subject in the process of preparing to write a report or proposal, you also learn more about a subject in the process of preparing to speak about it to others. View each speech as a chance to learn as much during the preparation process as you want your audience to learn during its delivery.

In short, improving your speaking skills first requires you to focus your attention and discard negative thoughts. Then you can proceed to follow the guidelines in this book for preparing and delivering your best speech. Adopting a positive attitude and working hard go hand in hand.

IMPORTANCE OF THIS BOOK TO YOU

We've established that the act of preparing and delivering speeches requires a great attitude and hard work. It's only natural for you to ask, "Why are speeches worth all this effort?"

The answer is that success in your professional life will depend on your speaking skills—either because you regularly will be asked to speak before groups or because you will occasionally be asked to do so. In the first case, much of your job may involve presenting new ideas to colleagues in your own organization or speaking about products or services to your customers. In the second case, you may toil away for years before you get that first request to speak before supervisors, clients, or professional colleagues. In either scenario, your next promotion, your next job, or your professional credibility may depend on the skills emphasized in this book.

Besides helping you influence others, there's another equally important reason to use this book in refining your speaking skills. Every time you deliver a competent presentation you add to a reservoir of self-confidence that spills over into the rest of your life. Success in public speaking breeds success in interpersonal communication because similar skills are at work. Your ability to inform, persuade, and entertain in an oral presentation increases the likelihood that you will run a more effective meeting, give a better job interview, or respond more appropriately in a performance evaluation.

Thus your ability to speak effectively in public will enhance the quality of your personal and professional life. Now, how is this little book designed to help you quickly achieve the goal of effective public speaking?

PLAN OF THIS BOOK

A single design principle drives this book—namely, minimalism, which is sometimes defined as "less is more." It aims to give just enough guidance to help you with the five main challenges of your work: research, organization, text, graphics, and delivery.

Each chapter presents guidelines, supporting examples, and practical exercises. Absent are long explanations and complex theory, which help little with the work of preparing and delivering a speech. Specifically, the text includes the following chapters and appendix:

Chapter 1: Overview of Public Speaking

Chapter 2: Research

Chapter 3: Organization

Chapter 4: Text

Chapter 5: Graphics

Chapter 6: Delivery

Appendix: Sample Speech

The first chapter sets the scene by providing a summary of the process detailed in the rest of the book. It will help you choose what parts of the book to read and give you the basics needed to begin preparing speeches—even before you read the other chapters. Then the remaining chapters work through guidelines for completing each stage of the speech process. The appendix to this *Pocket Guide* contains a sample speech, along with marginal annotations that show how it follows the book's guidelines.

ACKNOWLEDGMENTS

First, I want to thank my editors at Prentice Hall, Steve Helba and Debbie Yarnell, for suggesting that I write this pocket guide. They also helped shape my three other Prentice Hall books: *Technical Writing: A Practical Approach* (4th edition, 2000; 5th edition, expected 2003); *Proposal Writing: The Art of Friendly and Winning Persuasion* (2000, co-authored with Charles Keller); and *Pocket Guide to Technical Writing* (2nd edition, 2001).

Special thanks go to Dick Hahn, my teaching colleague at Southern Polytechnic State University, who offered suggestions incorporated into this book. I have benefited from his experience as a former executive in industry, as an instructor of speech communication, and as a professional speaker. The other individuals who supplied ideas or text have my deep gratitude, especially Dory Ingram, Kim Meyer, Hattie Schumaker, Betty Seabolt, James Stephens, Shawn Tonner, and Stephen Vincent. These contributions first appeared in my previous books, from which I have borrowed here.

As always, I relied on my wife Evelyn, my son Zachary, and my daughter Katie for their constant support and especially for their help preparing the final manuscript. I must add—with not a little regret—that important things did not get said and done because of the time I spent on this and other writing projects. I appreciate their love and understanding more than I am able to express.

Contents

1 Overview of Public Speaking

You've probably heard bad speeches in your life—real clunkers from which you were glad to escape with your life. But no doubt you've also heard some excellent presentations—perhaps at a sports banquet, local lecture series, or political campaign. At the time you may have thought, "What a great speaker!" and wished that you, too, had the gift of public speaking.

Certainly some people take to giving speeches more naturally than do others. However, good speaking is much like good writing in that your ability to follow a disciplined process counts much more than native talent. In its simplest form, the process can be described in terms of the "3Ps":

*P*repare carefully

*P*ractice often

*P*erform with enthusiasm

This book rests on the belief that you can become a good public speaker—perhaps even a *great* one—if you abide by the guidelines presented here, all of which flow from the 3Ps. Sticking to them will prepare you to deliver a speech at school, on the job, or in your community.

This chapter includes the following sections that provide background information about public speaking:

- SLMR ("slimmer") Speech Model
- Speech Types: Based on Delivery Method
- Speech Types: Based on Purpose
- Five Elements in the Speech Process

The five "elements" are research, organization, text, graphics, and delivery. They form the titles, and thus the content, of the five chapters that follow this one.

SLMR ("SLIMMER") SPEECH MODEL

The SLMR (pronounced "slimmer") speech model presented in Figure 1–1 applies to all forms of speech events. As the acronym suggests, this model is indeed "slimmer" than the complex formulas sometimes put forth to describe speech communication. As well, it places more emphasis than do most models on the importance of listener feedback, both during and after the speech. Considered together, the four SLMR components—speaker, listener, message, and response—create the context for every speech you'll ever present. Following is an overview of each part.

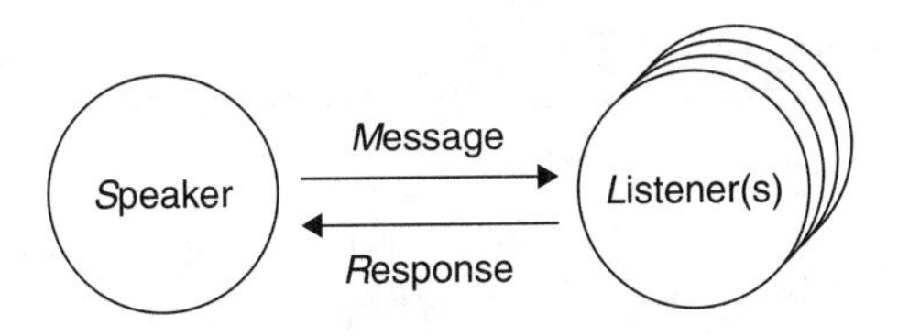

FIGURE 1–1 SLMR (slimmer) speech model.

SPEAKER

As speaker, you exert the most control over the communication context. Your words have the power to inform, persuade, or entertain. Your attitude, tone, and general demeanor can sway a listener in many different directions.

Of course, only in the ideal world would you as speaker exert total control over the speech event. In the "real" world, you often lose partial control for several reasons:

- The content and purpose of the speech may have been selected by someone else—for example, when a supervisor asks you to speak at a meeting.
- The medium may have features not of your choosing—for example, when a conference room has poor lighting or acoustics.
- The listeners may present you with a challenge—for example, when they have very little knowledge of your topic.

However, these challenges don't change one simple fact. You, as speaker, must give a successful speech in spite of the obstacles you face.

LISTENER

Like readers of written communication, listeners of oral presentations are your "boss," so to speak. Their needs drive every decision you make in planning and delivering a speech.

On first thought, you may think it's easier to convey information to a listening audience than to a reading audience. After all, there's no doubt that the audience is hearing—though, it should be noted, not necessarily listening to—your speech. With writing, you have no idea if readers have put down your written report after the first page. However, once you recall how your own mind drifts during boring speeches, you realize that listeners can "check out" just as quickly as readers who put down a document to read their e-mail or go to lunch. Following are some challenges you face in meeting the needs of your listening audience:

- They have diverse backgrounds.
- They may not want to be listening to a speech right now.
- They have short attention spans.

Chapter 2 gives more detail about researching the needs of your audience. For now it's worth noting that members of the same speech audience can differ markedly in their education, professional background, family history, culture, nationality, gender, and expectations for the speech.

Certainly public speaking presents no more important challenge than meeting the needs of listeners. In fact, just as the rule "Write for your readers, not for yourself" drives the writing process, the following rule lies at the heart of your preparation for every speech event:

Rule 1: Speak for your Listeners, Not for Yourself

As obvious as this directive may seem, it has been ignored by many a speaker.

MESSAGE

Every speech involves two-way communication. The part for which the speaker is responsible is called the "message," and it has two main parts: content and form. Content includes the information you deliver. Form includes the structure of the message, its style, and the features of your delivery.

Crafting an effective message requires considerable time, which must be used wisely. Many speakers end up spending too little time on the most important tasks (like outlining and practice) and too much time on less important ones (like fancy visual aids). Here are some essential tasks related to producing an effective message:

- Locating specific information about your listeners
- Shaping the speech around their expectations
- Following a simple structure for the text
- Practicing in a way that prepares you for the "real thing"
- Staying flexible enough to respond to feedback

If the preceding points seem like common sense, they are. Like a good piece of writing, a good speech embodies simplicity in design and delivery.

RESPONSE

Too often speeches are viewed as a one-way delivery of information. Although it's understood that the audience may ask questions, experts rarely place enough emphasis on (a) the reaction of listeners during and after the speech process and (b) the corresponding response of the speaker. That's why the speech model in Figure 1–1 gives equal treatment to communication that moves from speaker to listener, on the one hand, and from listener to speaker, on the other hand.

Listeners express their responses in three main ways: as nonverbal (occasionally verbal) behavior before and during the speech, as questions immediately after the speech, and as questions or comments received later. Two of three types can influence the speech itself, and the third can influence your next communication with the audience. Here are a few points to consider about reader response:

- Establish rapport before the speech
- Observe body language during the speech
- Adjust your speech in response to the audience
- Handle the question period as if it's a continuation of the speech
- Provide follow-up as closure

This approach to understanding and reacting to the response of the audience places great importance on the audience portion of the speech model.

The next part of this chapter's overview divides presentations by their four main modes of delivery.

SPEECH TYPES: BASED ON DELIVERY METHOD

In the first moments of a speech, listeners can observe what mode of delivery the speaker has selected. In this instant they form an immediate impression of the speaker and the type of experience they expect to have. If the speaker is reading a text from behind the lectern, for example, his or her speech will be received differently than if the speech were being delivered from a few notecards by a speaker wandering around the room.

Here we'll examine the four methods of delivery: verbatim, memorized, extemporaneous, and impromptu. Each section that follows defines the term, gives a context for its use, and notes any important advantages or disadvantages. The preferred method of delivery for most speeches is extemporaneous.

VERBATIM

The word "verbatim" comes from Latin and means "using exactly the same words." In this form of delivery you are reading a speech word for word, either from hard copy or from a teleprompter screen.

Because few people other than television broadcasters or politicians use a teleprompter, we'll focus on the use of hard copy. Speakers sometimes prefer this mode of delivery in the following situations:

- When they believe the exact phrasing of the speech is crucial to its effect on the audience
- When they're concerned they may speak too long if they don't have a prescribed text
- When they like the "comfort zone" of a prepared presentation whereby they only have to read words on a page, not "think on their feet"

A verbatim speech rarely is the right choice, but there are exceptions. One might be a lecture by a well-known person, who also is an excellent writer. The speaker might be able to deliver a speech or academic paper while still maintaining eye contact with the audience and creating an animated delivery. For most speakers, however, it's a bad idea to read a speech. Too often the delivery turns stiff and overly formal with little eye contact with the audience. Feeling ignored by a speaker who drones on, listeners will daydream and finally mentally depart altogether. In fact, this writer has seen people so put off by a verbatim speech that they *physically* departed from the scene. It's not clear that the speaker, with eyes lowered and nose in the speech text, even noticed.

Memorized

Like a verbatim speech, a memorized speech usually starts with a text having been written out word for word. Then the speech is practiced to the point that it is committed totally to memory such that notes or outlines are not used.

Speakers with good memories sometimes choose this format for the following reasons:

- They believe it reduces the likelihood they will digress from the topic.
- They think memorizing the speech shows the audience that they have prepared well.
- They believe they will be able to focus more on delivery when they are not dealing with an outline or notes.

Ironically, memorized speeches work best if the speaker looks down at some notes even when they are not needed or used. This technique makes the speaker seem less wooden. However, it's best to avoid memorized speeches altogether. Besides making you seem too stiff and formal, they can go awry if you forget part of the speech or lose your place after getting interrupted during delivery.

Extemporaneous

The term *extemporaneous* (or "extemp") refers to a speech that is delivered from notes. Though thoroughly familiar with the material, the speaker does not commit the speech to memory and instead develops final wording during the speech.

Extemp speeches are by far the most common format in college classes, in business, and during professional conferences. They are preferred for the following reasons:

- The speaker appears to be more natural and less formal.
- The perceived informality reduces the distance between speaker and audience.
- The speech can be altered to respond to changing circumstances during the speech.

The most powerful argument for choosing extemp delivery is the third point. You communicate best with listeners when your mind stays active and responsive to changing conditions. An audience can sense the sincerity in such a presentation and appreciates the effort it takes.

IMPROMPTU

An impromptu speech is one delivered on the spot with no serious preparation. Rarely used in formal settings, it usually occurs when you feel compelled to rise to speak on an issue at a meeting.

Some speakers handle the impromptu format better than others do. They simply may perform better when called upon to express an opinion without any preparation. Yet all speakers stand a better chance of delivering good impromptu speeches if they master the techniques of extemporaneous speaking emphasized in this book. Most of the delivery techniques are the same.

The next section moves from different methods of delivery to different purposes for giving a speech.

SPEECH TYPES: BASED ON PURPOSE

As a training expert at your firm, assume you've been asked to speak at the firm's annual board meeting about advances in distance education—especially Internet courses. Like many speeches, this one incorporates aspects of all three main purposes for oral presentations: to inform, to persuade, and to entertain. You are informing the board about recent advances, you are persuading them you have credibility to speak on the topic, and you had better be entertaining them to some degree because they will be sitting through eight other speeches at that meeting.

Despite its multiple purposes, your speech to the board falls mainly into the category of an informative speech. Likewise, in most speeches one of the following three purposes predominates: informing, persuading, or entertaining. Brief descriptions of each type follow, along with examples from the working world.

Informative Speeches

The chapter 2 discussion on audience divides listeners into several main groups, of which "decision makers" are the most important. They usually want information that will move them closer to making up their mind on an issue. In other words, they may just want information from you, not argument or strong opinion. Many of the speeches that you give in school and in your career will take this form. Following is an overview of the basic pattern for an informative speech and a brief example of a situation that requires one.

Chapter 3 (Organization) presents an organizational pattern that applies to all speeches, and chapter 4 (Text) details specific patterns used in informative speeches. For the purposes of the brief overview here, following are the main patterns used to develop the text of an informative speech:

1. Definition
2. Description
3. Classification/Division
4. Comparison/Contrast

The following example would allow you to employ several of these techniques.

As an example of the context for an informative speech, assume your employer has decided to open an office in Tokyo, Japan. As a member of the human resources staff, you have been asked to give a speech describing features of Japanese culture to employees who will be visiting Japan. These employees will be opening the office, hiring local employees, and meeting with Japanese clients. In presenting highlights of Japanese culture, you will be contrasting Japanese and U.S. formalities in writing, speaking, and interpersonal communication. The example in the next section uses this same framework to describe the scenario for a persuasive speech.

Persuasive Speeches

Although many situations require an informative speech, other times listeners expect to hear opinions. In the working world, for example, you may be the technical expert asked to give your views on a proposed change, such as a shift to a health maintenance organization within your organization. In the academic world, your instructor may ask you to present your solution to a problem on campus or to "take a stand" on an issue. In each case, your aim is to persuade the audience to accept your views. This section describes the basic pattern for a persuasive speech and describes a situation that would call for such a presentation.

Chapter 3 (Organization) provides a pattern that applies to all speeches, and chapter 4 (Text) gives suggestions for writing persuasively. Listed here are some of the more common techniques for developing the body of a persuasive speech:

1. Using evidence correctly
2. Choosing the most convincing order for points
3. Being logical
4. Citing only appropriate authorities
5. Avoiding argumentative fallacies
6. Refuting opposing arguments

The case study that follows would require you to use various argumentative techniques, in combination with some informative ones. It involves a context similar to one used earlier for an informative speech.

Assume your company produces health products for hospitals and is considering whether to open a sales office in Tokyo, Japan. Your sales to Japan have grown steadily for ten years, and you are spending considerable money sending sales representatives to Japan to meet with customers. As a member of the human resources staff, you've been asked to analyze the data and recommend for or against starting the office. Having decided to recommend the office in your speech, you will (1) contrast the cost of sending staff to Japan for the next five years versus the cost of operating an office in Japan for that same period and (2) give four main reasons for hiring a native Japanese person to run the office, rather than a U.S. employee with experience in Japan. The example in the next section uses this same situation to examine the context for an occasional speech.

Occasional Speeches

Besides speeches that inform and persuade, a third category includes speeches that have a function that is often just as important—to engage the attention of listeners at a special event. One could rightly argue that all speeches must engage the listeners' attention or risk not keeping them interested enough to hear what is being presented. However, for some speeches this purpose is the main goal. Examples include introductions to other speakers, keynote speeches at meetings, and kick-off speeches for seminars. This section gives an overview of some simple techniques to use in such occasional speeches and presents a case study that involves such a speech.

Listeners have high expectations for presentations that aim to entertain, both because they know material will be light and also because these sorts of speeches often occur at the beginning of events when their attention is high. Chapter 4 (Text) will provide detailed guidelines for writing the speech text. For our purposes here, following are a few techniques that apply to any speech that aims to entertain an audience:

1. Emphasize narrative—that is, stories—over explanations
2. Use humor with care, making sure to gauge your audience well
3. Use gestures and other body language liberally
4. Choose the appropriate length
5. Avoid digressions that can derail an occasional speech

The following context for an occasional speech is the same as that used for the previous examples for informative and persuasive speeches.

Assume that your firm has decided to open an office in Tokyo, now that business prospects in Japan appear to justify such an investment. As the salesperson in charge of the Japanese market for the last five years, you've worked with Japanese customers without benefit of an office and have traveled to Japan several dozen times. For a banquet at the company's annual meeting, you have been asked to present an after-dinner speech about your experiences in Japan. The management believes some light remarks will be appropriate considering the decision to open a Tokyo office. You've decided to tell a few humorous stories about inadvertent mistakes you have made in adjusting to Japanese culture, in hopes of helping employees who follow you to Japan. Though mainly entertaining, your stories will remind colleagues to bow appropriately, use business cards often, and have their own interpreter rather than relying on the customer's.

FIVE ELEMENTS IN THE SPEECH PROCESS

Much like good writing, good speaking results from a rigorous process that starts long before the speech is given. The process comprises five elements: research, organization, text, graphics, and delivery. The overview that follows serves as a foundation for the rest of the book.

RESEARCH

Research is most often associated with the effort to find information on a topic already selected and to incorporate this borrowed information into your speech. However, if you have not been given a topic, some preliminary research must be performed just to find one. As well, it is important to conduct research on the background and preferences of your listeners. Chapter 2 (Research) covers all parts of the research process and has these main sections:

- Finding information about the audience
- Finding information about the topic
- Using borrowed information

Careful research is the essential foundation for your speech. Errors at the research stage can create big problems for you later, so devote plenty of time to getting the information you need at the outset.

ORGANIZATION

After securing useful information about your audience and topic, the next task is to organize it in a fashion that accomplishes the purpose of your speech. Chapter 3 (Organization) presents a speech format that has three main parts: Abstract, Body, and Conclusion (ABC). The chapter also shows you how to use outlines to organize material easily and appropriately. The chapter's main sections are as follows:

1. Background on the ABC format
2. ABC format
3. Outlines

Effective organization may be even more important in a speech than it is in writing. Because listeners hear the speech only one time through, you must structure it so they get each point the first time they hear it.

TEXT

With outline in hand, sometimes you prepare notes from which you will speak. Other times, you write out the full text that you later convert into notes for the presentation. Writing out the text helps establish a unified flow of ideas in much the same way that creating multiple drafts accomplishes the same purpose in a document. Chapter 4 (Text) will assist you by providing suggestions for writing the speech draft. It will give guidelines for patterns of organization used in informative, persuasive, and occasional speeches. Thus the chapter's main sections are as follows:

1. Text patterns for informative speeches
2. Text patterns for persuasive speeches
3. Text patterns for occasional speeches

Writing the speech text also requires careful editing for grammar and style. First, the text of the speech may be presented in print at a later time. Second, focusing on careful editing will help eliminate the kinds of grammatical errors, wordiness, and technical jargon that bother listeners and damage your credibility. Although there's no room in this short book to cover these topics, you should consult a grammar and style manual for quick reference as you edit.

GRAPHICS

As you develop the speech text, you should be looking for opportunities to insert graphics into the presentation. Graphics enhance the speech by reinforcing main points, simplifying ideas, and creating interest. Chapter 5 (Graphics) includes the following main sections to help you develop graphics for any speech:

1. General guidelines for speech graphics
2. Specific guidelines for eight graphics
3. Misuse of graphics

No doubt, listeners expect graphics to accompany most oral presentations. However, the fact that we can now create sophisticated graphics on computers does not mean all listeners prefer them. In fact, simple visuals like overhead transparencies still work well in many contexts. Chapter 5 will help you choose, create, and use the most appropriate visual aids for each presentation.

DELIVERY

The previous parts of the process—research, organization, text, and graphics—prepare you for the challenge of delivering the speech. Chapter 6 (Delivery) provides detailed guidelines related to practicing and giving a presentation. As such it has the following main sections:

1. Techniques for practice
2. Guidelines for delivery
3. Guidelines for answering questions
4. Guidelines for handling anxiety

Although you will develop your own individual style in speaking just as you have done so in writing, some basic rules apply to most speakers in most situations. The fact is that effective delivery largely determines the success of your speech.

CHAPTER SUMMARY

This chapter introduces you to the subject of giving good speeches by covering four main topics. First, it presents the SLMR ("slimmer") speech model that includes a four-part structure—speaker, listener, message, and response—for understanding the context of every speech. Second, it classifies speeches by four modes of delivery: verbatim, memorized, extemporaneous, and impromptu. Extemporaneous speeches are the most effective because they rely on solid preparation without appearing to be stilted. Third, it uses the criterion of purpose to classify speeches into those that are informative, persuasive, or occasional. Fourth, it summarizes the five parts of the speech process that are detailed in the next five chapters on research, organization, text, graphics, and delivery.

EXERCISES

Exercises 1–5 have several options from which you can select for completing the assignments:

- **Written or oral response:** All exercises can be completed by preparing either a short paper or a short speech.
- **Individual or group response:** All exercises can be completed as an individual assignment or as a group project.

Your instructor will indicate the criteria you should follow in completing the exercises that follow.

1. **SLMR Speech Analysis—Text of Another's Speech**
   The purpose of this assignment is mainly descriptive, not judgmental. Select a written speech example from a textbook (such as a freshman English reader), from a periodical (such as *Vital Speeches*), or from another source. Describe the speaker, listeners, and message expressed in the speech. In addition, describe what you think the audience response might have been and why. If your instructor requests, submit a copy of the speech.
2. **SLMR Speech Analysis—Videotape of Another's Speech**
   The purpose for this assignment is descriptive, not judgmental. Use a videotape of a speech from your library, a videotape that you made of a speech, or a videotaped speech provided by your instructor. Describe features of the speaker, listeners, message, and response. At the discretion of your instructor, be prepared to submit the videotape.

3. **SLMR Speech Analysis—Your Own Speech**
   Select a topic for a speech you could give that would fit one of the three categories in this chapter: informative, persuasive, or occasional. After some reflective thinking about what you hope would be your experience preparing and delivering the speech, describe features of the speaker, listeners, message, and response. This exercise requires that you speculate on *intended* consequences—that is, what you would like to see happen.
4. **Steps in the Speech Process—Critical Analysis**
   The purpose of this assignment is both descriptive and judgmental. Using the text or the videotape of a speech given by someone else (for possible sources see Exercises 1 and 2), analyze the quality of the speech from the standpoint of the five elements in the speech process described in this chapter. In other words, explain the degree to which you think the speaker succeeded in following the process. This assignment asks that you use only the limited information on the speech process available in this chapter, not the more detailed information included in later chapters.
5. **Interview of Person Who Gives Speeches**
   Interview someone who you know has given any kind of speech within the last year. Report on the individual's answers to some or all of the following questions:
   a. What did the audience expect of the speaker?
   b. What was the background of the audience?
   c. What was the purpose of the message delivered?
   d. What was the response of the listeners?
   e. Was the speaker pleased with the result? Why or why not?
   f. What suggestions does this speaker have for individuals who are attempting to acquire the skills of effective public speaking?

2 Research

Effective speaking requires effective preparation. If you have prepared adequately, you are much more likely to deliver a speech your audience remembers for all the right reasons. An important part of preparation involves research, which is the subject of this chapter. Three main research tasks are covered here:

- Finding information about the audience
- Finding information about the topic
- Using borrowed information

These three tasks are quite different in what they require of you. In the first, you examine the nature of your audience and then plot a strategy to meet its needs. In the second, you conduct a two-fold research activity of generating a topic and then finding information to support it. In the third, you make certain to distinguish borrowed information from your own material, especially when the text of the speech is written out.

More specifically, the first section of the chapter lists problems in audience analysis, describes the technical and decision-making levels of listeners, and then suggests ways to meet their needs. The second section offers guidelines for choosing a topic from personal or outside sources and then provides details about using three sources of research material: online catalogs, the library, and interviews. The third section offers guidelines for moving from source to notecards, to outline, and to draft in preparing a formal speech that relies on borrowed information.

FINDING INFORMATION ABOUT THE AUDIENCE

Every speech should aim to satisfy expectations of the listeners. As obvious as this goal may seem, it remains a deceptively difficult one to achieve. The key is understanding the problems that listeners face, researching their technical background, researching their decision-making roles, and following a systematic procedure for recording their needs. This section provides information and guidelines on these objectives.

UNDERSTANDING THE AUDIENCE PROBLEM

You're preparing a speech for a New Orleans trade show your boss asked you to attend. Before flying off to the show, you have written a draft and practiced it several times so you can give it from notes. You feel well prepared, though a bit nervous. Just before you leave for the airport, you run into an experienced marketing person in the office who reminds you that the trade show occurs during a festival week in New Orleans. He also notes that the trade show attracts customers who know little about the sophisticated products your company sells. This information introduces some variables you had not yet considered. Following is a list of obstacles faced by most speakers.

Obstacle 1: Listeners Are Always Distracted

Listeners are vulnerable to a variety of distractions. Some of them are external, such as noise from the room next door or problems with the ventilation system in the room where the speech is being given. Others are internal, such as the tendency all of us have to daydream while we're listening to a speech.

Obstacle 2: Listeners Are Impatient

Many listeners lose patience with speeches that are hard to follow, given in a monotonous tone, or last too long. They are constantly asking themselves questions such as the following:

"What's the point of this speech?"

"What does this topic have to do with my life or my job?"

"Where's the speaker going with the topic?"

"When will this end?"

Indeed, they've heard so many disappointing presentations that they may not have high expectations for yours. You need to acknowledge to yourself that most listeners are impatient so that you can prepare a plan to meet their needs.

Obstacle 3: Listeners Lack Your Technical Knowledge

In college speech courses—and even more so in your career—you will speak to audiences who lack the background you might have as a result of your education or personal knowledge of the topic. Although they may place value on a technically sophisticated speech, they want it delivered in language they understand. You should always avoid "talking over their heads." Think of yourself as an educator. If listeners don't learn from your presentation, you have failed.

Obstacle 4: Most Speeches Have Listeners with Diverse Backgrounds

If the entire audience for a particular speech always came from the same background, public speaking would be a much easier proposition than it is. A speech could be neatly tailored to the expectations of one like-minded group. Unfortunately, in the real world of public speaking, each audience comprises listeners with different backgrounds, different expectations, and different levels of authority. For example, an audience for a presentation about changes in company benefits might include the company president, secretaries, engineers, and sales staff.

Now let's move from the obstacles associated with meeting the listeners' needs to a strategy for overcoming them. The first goal is to realize that an audience may include individuals with varied technical backgrounds.

Sorting by Technical Level

Those listening to a speech probably differ among themselves in technical knowledge of the subject. Although the word "technical" often refers to disciplines such as engineering and technology, it has a broader meaning here. The most "technical" listeners are those who by training, experience, or interest have command of the technical information about a speech topic. Using the criterion of technical background, we can separate listeners into four main levels: managers, experts, operators, and general listeners.

Technical Level 1: Managers

Many employees with technical experience aspire to become managers. Once in management, however, they often become removed from the hands-on technical aspects of their professions. Instead, they manage people, set budgets, and make decisions of all kinds. You should assume that members of your audience who are managers might not be familiar with technical features of the subject. They often need the following:

- Background information
- Definitions of technical terms
- Grouping of similar points
- Clear statements of what should happen next

Technical Level 2: Experts

Experts include anyone with a solid technical understanding of the topic. These listeners may have many years of higher education—as with engineers and scientists—but that is not necessarily the case. For example, a custodial supervisor with no college experience would be considered an "expert" listener for a speech that recommends changes in the maintenance program for a company's physical plant. Whatever their educational levels, most experts need the following in a speech:

- Thorough explanations of technical implications of the topic
- Visual aids that include relevant supporting data
- References to sources used for technical support

Technical Level 3: Operators

Managers and experts tend to receive the most attention because they often make many of the decisions related to the speech. However, an audience also may include operators, who will put ideas in the speech into practice. Operators expect the following:

- Clear organization of the material
- Definitions of technical terms
- Clear connections between the speech content and their jobs

Technical Level 4: General Listeners

Also called "laypersons," general listeners sometimes possess the least amount of technical knowledge about the speech topic. For example, a speech about the environmental impact of a proposed landfill may be heard by many homeowners who have little or no biochemical knowledge of the toxic wastes associated with landfills. These general listeners often need the following in a speech:

- Clear definitions of technical terms
- Frequent use of graphics
- Distinction between facts and opinions
- References to the effect the speech has on their lives

As you see, there is some overlap among the expectations of all four technical levels. Yet there are also some major differences that suggest the need for careful research into the technical backgrounds of listeners. Another important source of information is the degree to which listeners have the authority to make decisions based on what they hear in a speech.

Sorting by Decision-Making Level

Besides being categorized by technical level, audience members can be classified by degree of decision-making authority. You should pay special attention to decision makers—that is, the listeners who are most likely to create change as a result of hearing your presentation. Consider the following three levels as you research the decision-making responsibility of your listeners.

First-Level Audience: Decision Makers

The first-level audience can act on the information you present. If you are proposing a new computer lab for your office, for example, first-level listeners will decide whether to accept or reject the idea. Or if you are proposing that students be given an opportunity to evaluate faculty, first-level listeners will decide on the practicality of your proposal. In other words, decision makers translate information into action.

Second-Level Audience: Advisers

This second group also can be called "influencers." Although they may not make decisions themselves, they give advice to decision makers on the basis of hearing a speech. For example, the second-level audience for a speech on a proposed change in course registration at a college may comprise experts such as the registrar and budget officer, who would advise the vice president for business operations about the feasibility of suggestions. Sometimes the most important listeners may be advisers to whom decision makers have delegated considerable authority.

Third-Level Audience: Receivers

Some listeners are not part of the decision-making process. Instead, they only receive information in the speech and make adjustments in their lives accordingly. For example, a speech to the sales department on procedures for handling Internet sales may be heard mainly by people who are expected to put the new procedures into practice.

Once you realize listeners come from varied technical and decision-making levels, you can begin to gather research that leads to a complete analysis of their needs. The next section gives guidelines for such a procedure.

GAUGING THE AUDIENCE

Knowing the challenges in analyzing your audience, you can begin to gather the kind of information about listeners that will help you design the speech. Use the following procedure, which includes completing forms in Figure 2–1 (Audience Matrix Form) and Figure 2–2 (Audience Preference Form).

Audience Guideline 1: Group Listeners by their Background

Before preparing the speech you must determine just how much diversity your listeners encompass. Use the grid in Figure 2–1 (Audience Matrix Form) to list either all of your listeners (for a small audience) or a sample of your listeners (for a larger audience). The grid identifies listeners by their technical level on the vertical axis and by their decision-making level on the horizontal axis. After completing the form, you will have a good sense of how the range of audience background will shape the speech.

Audience Guideline 2: List Details about Some or All of the Audience

This next step can be completed for all listeners or for a sample, again depending on the size of the audience. Realistically, most audiences are too large for you to describe the background of each listener. Instead you can select a sample of those to whom you are speaking. Use the form in Figure 2–2 (Audience Preference Form) to characterize features of these individuals. As the form indicates, your audience research aims to cover the following topics:

- Technical background
- Educational background
- Interest in the topic (high, medium, low)

- Main question he or she needs answered
- Main action, if any, you want him or her to take
- Personality features that affect his or her response to the speech
- General preferences in speeches (style, format, length, etc.)

Audience Guideline 3: Talk with Others Who Have Spoken to the Same Group

The previous two guidelines tell you *what* to search for in your audience research. This one and the next tell you *how* to find the information. Often your best source is a work colleague. Ask around the organization to find out who else may have spoken to the same or a similar audience. Useful information could be as close as the next office.

Audience Guideline 4: Locate Information on the Web

If you're speaking to individuals who work in an organization or belong to a professional association, you can often find biographical information on Web sites of the company, a professional association, or individual audience members. Such sites may describe educational background, work history, professional interests, and more—all details that prove useful in tailoring a speech to the needs of the audience.

Audience Guideline 5: Focus on the Needs of Decision Makers

The more detail you have about all members of the audience, the better you can plan a responsive speech. However, remember that your main focus should be on the needs of those who will act on information in your speech. This subset of the audience should get your special attention in speeches, just as it does in your writing.

Audience Guideline 6: Remember that All Listeners Prefer Simplicity

Even if you cannot find much information about the background of individual listeners, you can always return to a basic premise for all speeches: Listeners prefer short, easy-to-understand presentations. The popular KISS principle (Keep it Short and Simple) applies to speeches just as it does to writing. Few, if any, listeners ever complain because a speech is too easy to understand or too short.

FINDING INFORMATION ABOUT THE TOPIC

Just as with documents, research for speeches may lead you into familiar territory: the library. Yet speech research also can rely on information gathered from many other sources, such as personal interviews. This section first

Technical Level	Decision-Making Level		
	Decision Makers	Advisers	Receivers
Managers			
Experts			
Operators			
General readers			

FIGURE 2–1
Audience matrix form.

offers suggestions for conducting research to help you find a topic—in those cases in which you are permitted to choose your own. The rest of the section examines three main sources for researching information once you have selected a topic: online catalogs, the library, and interviews.

CHOOSING A TOPIC

Often you will have been given a topic by your college instructor or by your supervisor. However, there may be other times when you have the opportunity to develop a topic totally on your own (such as for a speech course) or to select one from within a subject field (for a conference related to your profession). Some research may be needed for this process. In this context, "research" is defined quite generally as the process of locating information that might help generate a topic, either from your own background or from sources external to you. Following are some guidelines for finding a topic for a speech.

Topic Guideline 1: Review Your Life Experience

You speak best about what you know well. Personal experience provides a wealth of information from which you can draw for presentations. Informative, persuasive, and occasional speeches alike can flow from the events of your life. Here are a few general subjects to examine in the process of reaching a narrower topic on which you will speak:

- Skills learned at summer jobs or at work during the school year
- Interesting characters encountered on the job

Directions: Give details on the following topics about the audience member listed.
Name and title of audience member:

____

____

1. Technical background

2. Educational background

3. Interest in the topic (high, medium, low)

4. Main question he or she needs answered

5. Main action, if any, you want him or her to take

6. Personality features that affect his or her response to the speech

7. General preferences in speeches (style, format, length, etc.)

FIGURE 2–2
Audience preference form.

- Close calls in your life
- Some "first time" experiences that taught you lessons
- Volunteer work that changed your life
- Travel experiences that taught you lessons in multiculturalism
- Characters in your family history, recent or long ago

Topic Guideline 2: Consider Hobbies or Special Interests

As already noted, good topics are those about which you have personal knowledge. If you look carefully at how you have spent your spare time, you probably will find a topic that will interest others—*if* you can show why it interested you. The trick in speaking about hobbies is conveying to an audience that same sense of wonder and enthusiasm you felt when you first took up the activity. Here are a few hobby-related subjects that might be a source of a speech topic:

- Your musical preferences and how they may have changed
- A favorite book or type of book and why it appeals to you
- An historical period that captured your interest and why
- A sports hobby and why it was or was not worth your time
- A highly technical hobby and how it challenged you

Topic Guideline 3: Chose an Academic or Career Interest

If you are in college, one natural choice for a speech topic would be your career or academic interest. Certainly this option is legitimate, but you need to be a bit cautious. First, you may not yet know enough about the field to speak about it knowledgeably. And second, there may be others in the audience—students and instructors—who know much more about the topic than you do. That said, it certainly is possible to give a good speech on your career interest, especially if you look for an interesting angle. Here are some possibilities:

- How a favorite hobby led to a career choice
- How a work experience formed the foundation for your career
- How a personal relationship shaped your career plans
- How a personal "calling" led to a career change
- How an "accidental" course selection led to a career choice
- How an internship led you toward, or away from, a career

Topic Guideline 4: Examine the Experiences of Others

When you don't have personal experience related to a topic or just want to avoid a self-focused speech, consider conducting an "oral history" project using the experiences of others. Many people can provide you with personal anecdotes, perspectives on historical events, views on current events, information about careers, and more. You may be surprised that even people you do not know you personally will be glad to assist you. Often such research sources for speeches are just waiting to be asked. Here are a few possibilities, some obvious and some not:

- Parents, grandparents, and other close relatives
- Close family friends
- Professionals in the career you have chosen
- Graduates of the college or university you attend
- Nonfaculty employees of your college or university

- Older citizens in retirement homes
- Recent or not-so-recent immigrants to the country

Topic Guideline 5: Explore Possibilities in Current Events
Current events provide excellent material for informative, persuasive, and occasional speeches of all kinds. Like speeches on career topics, however, topical speeches pose some risks you need to avoid. In particular, be sensitive to topics that might offend listeners or just plain irritate them. However, if your purpose is to choose a controversial topic, that's different. Also, you should be careful to choose a topic about which you can show some depth of knowledge, especially if there is any possibility the audience will include "experts." Here are few general subjects:

- Little-known information about a current event
- Contrasting views of a current event
- A current event that did not get much coverage but should have
- Your changing views of a current event over time
- An event you think will become historically important and why

Whatever topic you choose, be certain that (1) you can demonstrate enough knowledge to meet the expectations of your audience, (2) you have narrowed the topic so that it can be covered well in the time allotted, and (3) you believe the topic has a good chance of engaging the attention and interest of the listeners.

Using Online Catalogs

Both the library and the Internet can seem intimidating when you first start to gather research for a speech. Once you learn a few basics, however, you will gain confidence in your use of these amazing resources. This section includes some very basic information about using online catalogs to locate sources such as books and journals.

Most college and university libraries organize their collections by the Library of Congress (LC) system or by the Dewey Decimal system. The catalog is the roadmap to a library's collection of books, periodicals, and other material. As such it is an alphabetical list organized by author, title, and subject. Many libraries offer sophisticated online catalogs that can be searched remotely from home or office, as well as in the library itself. Unlike a manual catalog, the online version has additional features, such as keyword searching and information about the checkout status of a book. Also, the trend of providing access to the online catalog through the Internet means you can search the catalogs of other libraries and use links to other Web sites.

The rules for searching online catalogs vary depending on the computer program used by the library. The library's "help screen" is your best guide. Following are four of the most common strategies for finding information.

Online Guideline 1: Search by Author or Title

If you know the specific authors or titles of potentially useful books, conduct an author or title search to locate call numbers. Then study the catalog entry, especially the subject heading. Similar books can be found if you proceed to search by subject using these terms.

Online Guideline 2: Search by Subject Heading

Once you know the subject headings assigned to books and journals on your topic, searching by subject can be very efficient. Libraries select the subject headings from a set of books titled *The Library of Congress Subject Headings.* Knowing exactly which subject words to use can be a matter of trial and error. Once found, however, these headings can serve as powerful tools to gather information on the topic.

Online Guideline 3: Search by Keyword

This strategy is probably your best choice. It allows you to scan through all fields in a book's library record—author, title, subject headings, dates, notes, publishers, etc.—to locate books that match your request. Searching by keyword permits use of natural language instead of rigid subject headings. Pay close attention to the catalog's rules for keyword searches. You can often improve your results by limiting searches to particular fields.

Online Guideline 4: Switch between Keyword and Subject Searching

One of the easiest ways to find material on your topic is to switch between keyword and subject searching. For example, you can begin with a keyword search, selecting the books that best match your request. Then after examining the subject headings that describe these books, you can perform a subject search.

USING THE LIBRARY

At some point during your research you may need to visit your academic library. The library's services and collections of books, periodicals, microforms, and reference materials will support your research and help you locate additional information. Fortunately, most libraries are organized in a similar fashion, so the experience you gain using one can help in using another. This section offers guidelines for conducting library research.

Library Guideline 1: Take Advantage of Library Services

The three most important services the library provides are (1) reference and information, (2) interlibrary loans, and (3) circulation. Discussing your topic with a librarian can be an important first step and can save you considerable time. The librarian can recommend certain sources and guide you to sources, such as government publications, that you may not have found on your own. When you need sources outside the library where you are working, a librarian can order originals or photocopies from other libraries—sometimes for a fee and sometimes for free. And, of course, the circulation department helps you check out, renew, and recall materials that you may need.

Library Guideline 2: Use the Book Collection

As already noted, library catalogs today are usually automated—whether or not they can be accessed from outside the library. Books are arranged in an alphanumeric call number order in the library stacks, usually by Library of Congress number or Dewey Decimal number. Once you locate the exact book for which you are searching, browse through the books located beside this title. You will likely find other useful and related material.

Library Guideline 3: Use the Periodical Collection

Periodicals are publications issued at regular intervals. They comprise two main groups: popular magazines that take commercial advertising and professional journals that usually do not. Current issues, often housed near the circulation area, are arranged either alphabetically or according to subject. Back issues may be in the form of bound volumes of actual issues, microforms, CDs, or even full-text articles available through computerized indexes and databases.

Your key to using periodicals is an index or abstract. By searching for your subject in the index, you can find lists of articles, brief summaries in some cases, and even full text at times. Ask the librarian to advise you on the best index or database for your subject.

Library Guideline 4: Consult Newspapers for Topical Subjects

If your speech topic demands the most current information, newspapers provide an excellent source. One disadvantage, however, is that newspaper information has not "stood the test of time" to the same extent as information in journals and books. Another disadvantage is that most newspapers are either not indexed at all or are indexed only in their local area. Fortunately, there are print and online indexes for two of the best-known newspapers—*The New York Times* and *The Wall Street Journal.* Other online and Web-based options include *Lexis-Nexis,* a full-text index to some 5,000 publications, and *Newspaper Abstracts,* which indexes some thirty-two national and regional newspapers.

Library Guideline 5: Use Company Directories for Some Information

Often your speech topics may require that you find detailed information about specific firms. Many companies now produce sophisticated Web pages about their services and products. While not without bias, they can be a good source of information. Another, more objective source of information is company directories, some of which follow. You may want to ask a reference librarian to recommend others or to assist you in using the online versions of these databases.

- *D&B Million Dollar Directory*
- *Moody's Industrial Manuals*
- *Standard and Poor's Register of Corporations, Directors, and Executives*
- *Thomas' Register of American Manufacturers*
- *Ward's Business Directory of Largest U.S. Companies*

Library Guideline 6: Consult Dictionaries, Encyclopedias, and Other General References

Sometimes you may need general information on a speech topic. In this case you can consult dictionaries, encyclopedias, handbooks, and other sources. Some encyclopedias are even available as stand-alone CD-ROMs or as online or Web-based products. Although general references may help get you started on a project, the specialized references may be more useful. They often target a more scholarly audience, reflect greater knowledge of the subject, and refer to more in-depth materials in their bibliographies. A few examples follow:

- *Cambridge Encyclopedia of Life Sciences*
- *Dictionary of Business and Economics*
- *Dictionary of the History of Science*
- *Encyclopedia of Business Information Sources*
- *McGraw-Hill Encyclopedia of Science and Technology*

Although parts of this section have referred to use of the Web and the Internet, there is no room here to detail how the many Web search engines can help you locate Web sites on your topic. The Web presents exciting opportunities for researching your speech topics, but it also poses hazards. For example, the source and veracity of information often may be difficult to determine. Your best bet, once again, is to consult a reference librarian for helpful guidelines on using the Web as a resource for your speech.

Conducting Interviews

Some speeches require that you collect firsthand information, such as from personal interviews. Whether you are interviewing a family member or a person you have never met, follow clear-cut procedures so that you gather useful information. This section offers suggestions to prepare for and conduct a personal interview.

Interview Guideline 1: Express Objectives Clearly to the Interviewee

When you first contact the person to be interviewed, be precise about the purpose of the session. The conversation should (1) stress the value of the person's contribution, (2) put the person at ease about the general content of the questions, and (3) establish a clear starting and ending point for the upcoming interview.

Interview Guideline 2: Prepare an Interview Outline

Persons you interview understand your need for a written reference point for the interview. Indeed, they will expect it of any well-prepared interviewer. A written outline should include a sequential list of topics to be covered or questions you plan to ask.

Interview Guideline 3: Show You Value the Person's Time

You do this first by showing up a few minutes early so that the interview can begin on time. You also show this courtesy by staying on track and ending on time. Never go beyond your promised time limit unless it is absolutely clear the person being interviewed wants to extend the conversation further than planned.

Interview Guideline 4: Ask Mostly Open Questions

Open questions require your respondent to say something other than "yes," "no," or other short answers. They are useful to the speaker because they offer an opportunity to clarify an opinion or fact. They also are useful to you because you get the chance to gather more information.

Interview Guideline 5: Ask Closed Questions When You Need Specifics

A closed question is useful when, for example, you need a clear commitment from the person being interviewed. Perhaps the person has just given a complicated picture of a company's future in international marketing. To add clarity to the discussion, you might want to ask, "Do you think international sales will be a *higher* or *lower* percentage of your total sales in three years, compared to right now?"

Interview Guideline 6: Use Summaries Throughout the Interview

Frequent summaries serve as important, brief resting points during the conversation. They give you the chance to make sure you understand the answers that have been given. They also give your interviewee the chance to amplify or correct previous comments.

By following these guidelines and then taking good notes during the interview, you will secure useful details for your speech. Some of the most powerful supporting sources in a speech are the observations of people with whom you have made personal contact.

USING BORROWED INFORMATION

Many speeches do not depend on published information or other borrowed material. They are delivered in your own words from notes, involve no library work, and are not reproduced as text with notes and bibliographies. This part of the chapter doesn't apply to them. Yet other speeches do incorporate borrowed information. Therefore, they must subscribe to guidelines similar to those for documents that contain borrowed material.

Whether a speech is written out or not, you have an obligation to your audience to be clear about the degree to which you rely on borrowed information. Sometimes just a quick comment will do. Other times a more formal approach is needed. This section gives some background information on using sources correctly and then lists guidelines for doing so.

Avoiding Plagiarism

The following rule serves as the foundation for this book's suggestions on the use of borrowed information in oral presentations:

> With the exception of "common knowledge," you should acknowledge all borrowed information used in your speech.

Following is a rationale for using this rule in your speeches, as you use it in your writing.

"Common knowledge" is information generally available from several different sources in the field. When you are uncertain whether or not a piece of borrowed information is common knowledge, proceed to cite the source. It is better to err on the side of excessive citation than to leave out a reference and risk the charge of plagiarism. Plagiarism can be defined as the intentional or unintentional use of the ideas of others as your own. Here are three main reasons for documenting sources thoroughly, especially when speeches are written out in text:

1. **Ethics:** You have an ethical obligation to tell the audience where your ideas stop and where those of another person begin. Otherwise you would be parading the ideas of others as your own.
2. **Law:** You have a legal obligation to acknowledge information borrowed from a published source. In fact, generally you should seek written permission to use borrowed information that is copyrighted if you plan to publish your speech.
3. **Courtesy:** You owe the audience the courtesy of citing sources for those who may wish to find additional information on the subject later.

Certainly some plagiarism occurs when unscrupulous speakers and writers intentionally steal the work of others. However, most plagiarism results from mistakes during the research and drafting process, not from ill intent. The next section offers guidelines for avoiding such errors.

Tracking Borrowed Material

Whether you write out the text of your speech or just produce notes beforehand, you should take care to separate borrowed information from your own ideas. This chapter suggests that you follow a somewhat traditional—but still quite effective—procedure for avoiding plagiarism in speeches.

The documentation process described here assumes you may be writing out the text of the speech. Yet most of the steps apply even if you only produce speech notes or an outline. It further assumes that in producing speech text you would be following one of the many documentation systems included in handbooks such as the following:

- *Chicago Manual of Style*
- *Council of Biology Editors Style Manual*
- *MLA Handbook for Writers of Research Papers*
- *Publication Manual of the American Psychological Association*
- *U.S. Government Printing Office Style Manual*

No documentation system is presented in this *Pocket Guide* because space does not permit covering information readily available in writing handbooks. It is assumed that any speech you write out will include parenthetical references in the text to borrowed information and a list of sources at the end of the text—often called a "works cited," "references," or "bibliography" page.

Obviously, only a speech produced for distribution assumes the form of a documented paper with documented sources. For speeches that are given from notes, you can indicate your use of material that is not common knowledge by following these simple strategies:

- Tell the audience when you are shifting in and out of quoted material by using the words "quote" and "unquote."
- Give credit to sources that are paraphrased or summarized by a general remark before, during, or after the speech—whichever seems most appropriate.
- Prepare a list of "works cited" for speeches that depend heavily on outside sources, in case some member of the audience would find the list useful.

No matter what final form your speech takes, use the following guidelines when you may be collecting research other than common knowledge. Following is a procedure for transferring borrowed information from source, to notecards, to outline, and finally to draft.

Documentation Guideline 1: Write Complete Source Cards

Using 3″ × 5″ index cards, write a complete bibliography card for each source that you may use in the speech. Figure 2–3 shows a typical card and the sort of information it might include—author, title, publication information, library call number, and note to yourself at the bottom as to the potential usefulness of the source. Be precise in listing the information so that you will not have to consult the original source again when you assemble the final bibliography.

Documentation Guideline 2: Take Careful Notes on Cards

Most plagiarism occurs when notes are taken hastily. The important note-taking stage requires that you attend to detail and follow a rigorous procedure.

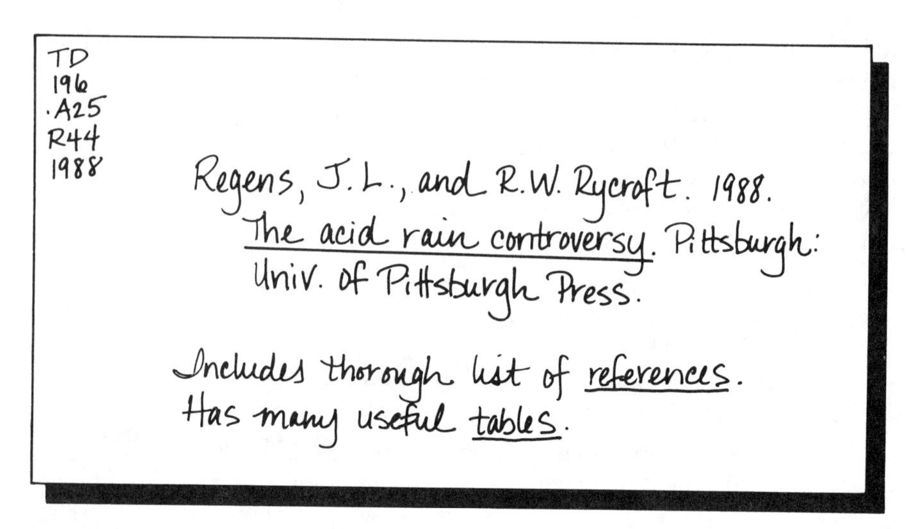

FIGURE 2–3
Sample bibliography card.

Among other things, you should (1) distinguish your own summary or paraphrase of a source from material directly quoted, (2) include exact wording of direct quotations, circling the quotation marks so that they will not be missed later, and (3) label the exact citation of the source on the source card (title, author, page) so that there will be no confusion later on.

The best way to avoid unintentional plagiarism at this stage is to divide your notecards into three basic types: summary cards, paraphrase cards, and quotation cards. Following are descriptions of the three types; Figure 2–4 includes examples related to a speech on the history of air pollution.

- **Summary cards** are written in your own words and reduce much borrowed information to a few sentences. They are best generated by reading a section of the source, looking away, and summarizing the passage in your own words. A summary card may include a few quoted passages, but the card's main purpose is to reduce much of the material to a brief summary. *As already noted, summary material must be acknowledged as being borrowed unless it is common knowledge.*
- **Paraphrase cards** include a close rephrasing of material from a source. They usually contain more of the source material than summary cards and thus demand more attention to the dangers of plagiarism. Like summary cards, they are best written by looking away from the source and rewriting the passage in your own words. You can use a few keywords from the source, but do not duplicate exact wording or sentence structure unless you integrate direct quotations into the paraphrase. *As already noted, paraphrased material must be acknowledged as being borrowed unless it is common knowledge.*

Quotation Card

pp. 5–6

"For at least the past two millennia, air pollution has been looked upon as a nuisance. As early as A.D. 61, the philosopher Seneca noted Rome's polluted vistas. Almost a thousand years later, the pollution associated with wood burning at Tutbury Castle in Nottingham was considered unendurable by Eleanor of Aquitaine, the wife of King Henry II of England, forcing her to move. Moreover, starting as early as 1273, a series of royal decrees were [sic] issued barring the combustion of coal in London in a futile attempt to address that city's burgeoning air quality problem. Such illustrations . . . //p6 underscore the enduring nature of air pollution as a public concern."

Quotation card records words exactly as they occur in the source. Ellipses (. . .) indicate some words are deleted. Hash marks (//) indicate a page change. "Sic" indicates a possible grammar error ("were" instead of "was") present in the source.

Paraphrase Card

pp. 5–6

Air pollution has plagued mankind for at least 2000 years. For example, Seneca commented on the polluted air of ancient Rome in A.D. 61. Also, Eleanor of Aquitaine, (wife of Henry II) was so bothered by the pollution at Tutbury Castle that she moved out. Although there were royal decrees about coal burning in London, the city continued to have major problems with dirty air.

Paraphrase card rephrases quoted passage above, including most of the main ideas.

FIGURE 2–4
Sample notecards from same passage.

- **Quotation cards** include only words taken directly from the source. Your main concern should be the exact transfer of information from source to card. As shown in Figure 2–4, you should use ellipses (spaced dots) when you delete words that you think are not necessary, but be sure that your deletions have not altered the meaning of the original passage. *All quoted material must be acknowledged as being borrowed.*

Summary Card

pp. 5–6

Air pollution has been a problem for a long time. In fact, it was noted by Seneca in A.D. 61 regarding ancient Rome.

Summary card presents only a brief overview of quoted passage.

FIGURE 2–4 (*Continued*)
Sample notecards from same passage.

Documentation Guideline 3: Transfer Notes Carefully from Cards to Draft

Unintentional plagiarism can occur when notes are transferred inaccurately from notecards to draft. As you have done on cards, circle quotation marks on the draft so that it is absolutely clear what is borrowed, on the one hand, and what is either your writing or paraphrased material, on the other. Again, remember that all borrowed information must be attributed to a source unless it is common knowledge.

CHAPTER SUMMARY

This chapter addresses several types of research related to speeches. First, it covers audience analysis. After listing obstacles such as the diverse background of people listening to the same speech, the chapter offers a strategy for characterizing listeners by technical and decision-making levels. The Audience Matrix Form and Audience Preferences Form are provided for that purpose. The next part of the chapter offers suggestions for finding a speech topic when you are not assigned one. Then you are introduced to the process of conducting research. Guidelines are included for using online catalogs, library sources, and personal interviews. Finally, the chapter recommends a procedure for handling borrowed information properly.

EXERCISES

1. **Audience Analysis—Your Class**
This exercise requires that you be separated into groups of three to six individuals. Assume you are about to give a speech to your class. Select a topic about which you might conceivably deliver a speech. First complete one Audience Matrix Form for the entire group, and then complete an individual Audience Preference Form for each member of the group.
2. **Audience Analysis—Your Workplace**
This assignment assumes you are preparing a speech on a subject dealing with a current job. Interview three to six individuals who could be members of the audience for such a speech. If possible, choose people with different technical and decision-making backgrounds. Complete one Audience Matrix Form for the entire group, and then complete an individual Audience Preference Form for each person in the group. As an option, you can choose a previous job as long as you are able to contact individuals in person from that organization.
3. **Topic Selection—Using the Five Guidelines**
This chapter includes the following five guidelines for choosing topics when they have not been assigned:

    - Review your life experience
    - Consider hobbies or special interests
    - Choose an academic or career interest
    - Examine the experience of others
    - Explore possibilities in current events

    At your instructor's discretion, use one or more of the guidelines to select five possible topics for each guideline used. Each topic should be specifically related to you and your experience, not to a generalized subject.
4. **Research—Online Catalogs**
Select a speech topic about which you believe there might be library research available. Then conduct a search for possible sources using an online catalog available to you at a source outside the library building. Use all four guidelines listed in this chapter for online catalogs, including the recommendation to switch between keyword and subject searching. At the discretion of your instructor, present the results of your search either as hard copy or as an attachment to an e-mail.
5. **Research—Library Resources**
Choose a speech topic about which you could find information in at least four of the five library sources mentioned in this chapter: (1) books,

(2) periodicals, (3) newspapers, (4) company directories, and (5) dictionaries, encyclopedias, and other general references. Report on the results of your search to your instructor, who will indicate the form your reporting is to take.

6. **Research—Interview**

   Option A: Team up with a member of your class. Conduct a focused interview based on questions you develop beforehand or based on other directions from your instructor. Present a written record of the interview and/or orally report results to the class.

   Option B: Interview a family member or friend, using a focused list of questions developed beforehand. Present a written record of the interview and/or orally report the results to the class.

7. **Using Borrowed Information—from Source to Notecards**

   Choose a topic about which you intend to give a speech and for which adequate library information is available. You may want to use a topic developed for Exercise 4 or 5. Using two or more sources that contain useful information, write at least ten notecards. Include at least two examples each of summary cards, paraphrase cards, and quotation cards.

3 Organization

One main feature separates written presentations from oral presentations. While perusing a written report, readers control the pace at which they read. They can navigate within the document at will. While listening to a speech, however, listeners are powerless to control the pace. They have no chance to "rewind the tape" if information is missed. This basic difference between reading documents and hearing speeches suggests that you must organize speeches very well. The audience should be able to absorb your message the first time through, for there may be no other opportunity.

To give listeners a structure for capturing information, this book recommends a three-part structure called the *ABC format*. As the diamond-shaped graphic in Figure 3–1 (ABC Format for Speeches) shows, the format has the following sections:

- Short introductory section called the *abstract*
- Long discussion section called the *body*
- Short wrap-up section called the *conclusion*

This beginning–middle–end format may look familiar to you. For example, it resembles the recommendation put forth in the well-known preacher's maxim, which goes something like this: "First you tell 'em what you're gonna tell 'em, then you tell 'em, and then you tell 'em what you told 'em." It also looks much like the structure of a typical written essay.

To help you understand the rationale for the ABC format, the next section of this chapter gives some background information. It is followed by a section that provides specific guidelines for developing the abstract, body, and conclusion. The final part of the chapter explains the rationale for using outlines and gives guidelines for doing so.

BACKGROUND ON THE ABC FORMAT

This section provides a rationale for the ABC format and describes its three sections: abstract (or introduction), body (or discussion), and conclusion (or wrap-up). The next section includes hands-on guidelines for choosing information appropriate for each of the three sections.

Well over 2,000 years ago, Aristotle wrote in the *Poetics* that the structure of a story forms its "basic principle, the heart and soul." He noted that plot structure must include three separate but integrated parts: the "beginning, middle, and end." It also must describe "an action which is complete" and "should neither begin nor end at any chance point" (*Aristotle Poetics,* translator, Gerald F. Else, Ann Arbor Paperbacks, 1970, pp. 28, 30). Although he was speaking about the plot of a tragedy, Aristotle was giving us guidelines that apply just as well to oral presentations.

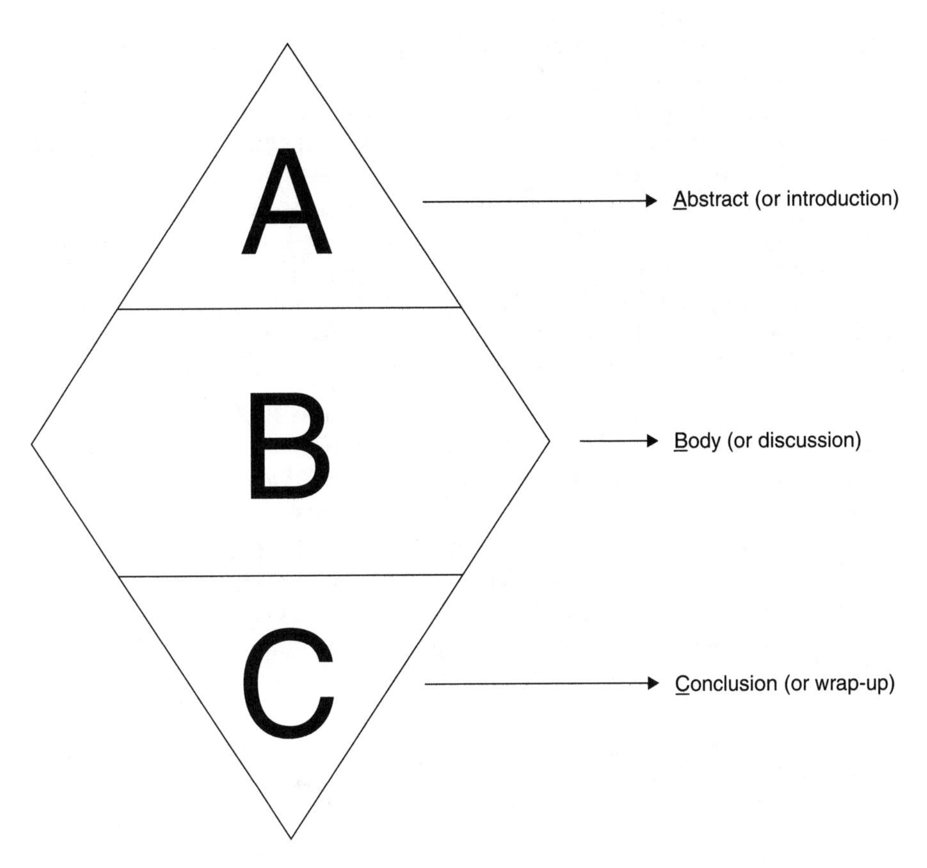

FIGURE 3–1
ABC format for speeches.

Aristotle's idea about a clear "beginning, middle, and end" serves to support the simple pattern used in this book. The ABC format rests on five assumptions about the preferences of listeners:

- They need a clear mental outline into which to place information provided in a speech.
- Their experience with diverse media—such as books, newspapers, film, and television—has shown them that most forms of communication work best when they have a clear beginning, middle, and end.
- The beginning of a work should explain why the speech is important and where it is going.
- The middle should follow through on the promise of the beginning by using techniques such as examples, transitions, and mini-summaries.
- The end should summarize the speech and make clear what, if anything, will happen next.

Certainly, long speeches might have many sections covering diverse subjects. Yet even they should retain a simple three-part structure that shapes the entire speech. In the same way that you may have three main folders on your computer desktop, which contain many embedded folders and files, you will have three main sections in every speech—whether it's a 5-minute overview at a meeting or a 90-minute detailed report on a yearlong project.

Although all three parts of the ABC format contain important information for the audience, some sections receive more attention than others because of the way that people listen. Figure 3–2 (Reader Interest Curve and ABC Format) shows the ABC pattern superimposed on a "reader interest curve." As the figure indicates, listeners have the following priorities:

First priority—Abstract: You can count on the listeners' greatest physical and mental attention at the outset of the speech. They are most ready to be informed, persuaded, or entertained. Use this opportunity to create a good first impression—don't squander the chance.

Second priority—Conclusion: You have the second highest attention of listeners at the end of the speech. They want the true sense of an ending, words that summarize what they are supposed to have heard. On a more basic level, they may be hoping for an inspiring close because they are just plain tired from sitting. In any case, give them a closing they will remember—don't just stop.

Third priority—Body: You have the listeners' lowest level of attentiveness in the middle of the speech. It's not that your listeners aren't lis-

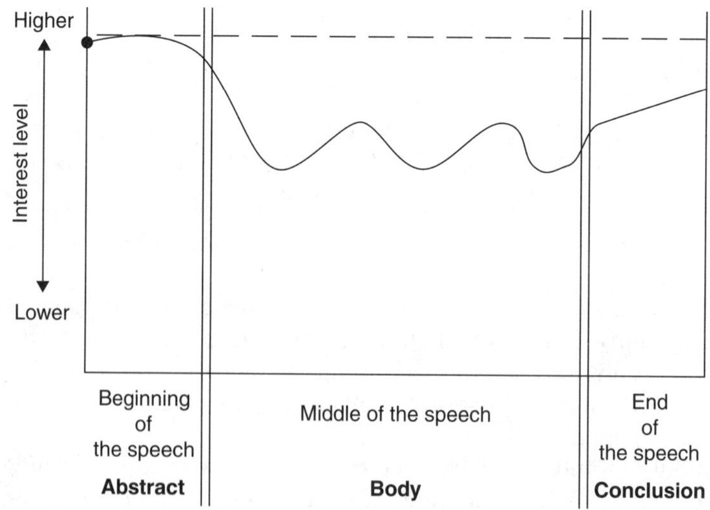

FIGURE 3–2
Reader interest curve and ABC format.

tening during the middle of the speech; it's just that they tend not to be as focused as they are at the beginning and end. This fact suggests you must keep their attention with rhetorical devices, such as changes in vocal tone or pauses in the speech.

These listening priorities should govern every aspect of your preparation for a speech. In fact, your focus on the needs of the audience is important enough to form the basis for the main rule governing every aspect of speech preparation and delivery. As already mentioned in Chapter 1, it is as follows:

Rule 1: Speak for your listeners, not for yourself

If you always remember this rule, you will design abstract, body, and conclusion sections that clearly meet the needs of your audience. Now let's examine specific rules for using the ABC format.

GUIDELINES FOR THE ABC FORMAT

In organizing your speeches, use the guidelines that follow to select material for the abstract, body, and conclusion.

ABSTRACT SECTION

The abstract provides an overview of the speech. As such it gives readers a sense of what should be considered most important. Every abstract should follow the **PIP formula—P**urpose, **I**mportance, **P**lan—as mentioned in the Preface. Here are the questions that need to be answered:

Purpose: What *purpose* does the speech have?

Importance: What *importance* does the speech have?

Plan: What *plan* will the speech follow?

The phrasing and order of the three PIP elements may vary from speech to speech, but almost every good speech includes all three. After you have presented the abstract, your listeners should be excited about the prospect of hearing what you have to say in the rest of the speech. Creating a sense of anticipation is your goal.

Abstract Guideline 1: State Your Purpose

Most important of all, listeners want to know why you are speaking. You may think the purpose is obvious. After all, doesn't the seminar program state that you are speaking on "Concerns of the Cuban-American Community in Miami"? However, the purpose is never so obvious that it should be omitted, for these reasons:

1. Making the purpose clear gives you a quick opportunity to establish a "comfort zone" for the reader. There will be no confusion as to what you intend to say.
2. Even if they've been told by others why you are speaking—as in a printed program or spoken introduction—listeners like to hear the purpose stated by you, the speaker. Their level of comfort expands as you confirm with your words what they have been told or what they have read.
3. The purpose statement sets a clear expectation and helps establish a good relationship with listeners. They now have the first criterion by which to judge the success of your speech.

Some speakers wonder just how obvious a purpose statement should be. The fact is that there's nothing at all wrong with beginning your speech with the phrase, "The purpose of this speech is to . . ." The examples that follow show several alternatives for phrasing the purpose.

Example 1: The purpose of this presentation is to describe diverse concerns of the Cuban-American community in Miami, Florida.

Example 2: This speech will describe diverse concerns of the Cuban-American community in Miami.

Example 3: In this speech I will describe three major concerns of the Cuban-American community in Miami, Florida.

Example 4: The Cuban-American community in Miami has many concerns related to its cultural background. Just what are these major concerns?

Example 5: Have you ever wondered what kinds of concerns have developed within the Cuban-American community in Miami?

These examples give options for either direct or more subtle statements of purpose. However, remember not to be *too* subtle about the purpose of the presentation. The listener should not have to work hard to find it.

Abstract Guideline 2: Show Importance

With purpose stated, you next must show the importance of the speech and generate interest. The amount of effort you expend on this step depends on the answers to the following questions:

1. Do listeners have a personal interest in the subject?
2. Does the topic relate specifically to their jobs?
3. Will the topic have an immediate effect on their lives?
4. Will the topic have a long-term effect on their lives?

The more "no" answers there are to these questions, the more likely it is that you must work to show the importance of the topic and create interest in it. Even when you know readers have interest, however, you still should spend at least a little time reminding them of why they should listen. For example, if you are speaking to a college class about the proposed increase in tuition, you might want to add to the students' already keen interest with a statement such as, "If the tuition increase is approved and you are currently a first-semester sophomore, you'll end up paying $3,200 more tuition before you graduate than you would pay under the current rules."

You can establish importance and create interest by techniques that have been used by speakers for centuries. Each is effective in the right context, but each can also seem out of place if used with the wrong group. As always, follow ***Rule 1: Speak for your listeners, not for yourself,*** by using techniques that fit your audience and context. With that caution in mind, here are some strategies for showing importance and creating interest:

Statistic: Enrollment at this university has increased 40 percent since January 2000, but the number of parking spaces for students has stayed the same.

Rhetorical question: As you reflect on your experience at this university over the last year or two, have you noticed that it's getting harder to find parking when you arrive for class?

Anecdote: Last week I spent 25 minutes looking for a parking space, which made me 10 minutes late for this class. That's one recent story, but I'll bet each of you has your own tale of woe about campus parking.

Startling statement: Faculty parking lots have been at least one-third empty every time I've observed them, but student lots are crammed full most mornings and early afternoons. Clearly, we have a parking problem on this campus.

Prediction: If this university is not able to respond to the problem of inadequate student parking, students may begin to transfer to nearby schools with better student services.

Credibility statement: For the last three years I've worked as a technical assistant for Rhodes Architects, Inc. Having assisted with several campus facilities design projects, I've developed some observations about the parking situation on our own campus.

Choose a method for showing importance that best suits the audience, the topic, and the speech situation. For example, the "startling statement" in the previous list might hurt your cause if you are speaking to a group of faculty and attempting to enlist support for building additional student lots. Find the right strategy for your specific listeners. Avoid an approach that might confuse, alienate, or bore them.

Abstract Guideline 3: Forecast Main Points of the Speech

The audience mainly uses only your spoken words to navigate through the speech. Therefore, you must use more obvious forecast statements than those used in written communication. Also called "organization statements," they become the mental outlines listeners need to follow in the rest of the presentation. Such statements also help *you* as the speaker. Once you've stated the main agenda, it will be easier for you to remember the wording and order of topics as you deliver the body of the speech. Following are suggestions for writing a good forecast statement:

1. Word points or topics exactly as they will be stated in the body of the speech, so that listeners can follow easily
2. List points in the forecast statement in the same order they will appear in the body of the speech
3. Consider using an overhead transparency or other visual aid for visual reinforcement of the points in your forecast statement

Handled well, the forecast statement of main points at the end of the abstract becomes a "roadmap" for listeners to use throughout the presentation. It is especially helpful for listeners who have let their attention drift. They can find their way back when you mention points they remember having been highlighted at the outset. Although such statements don't give much room for creativity, following are a few slight variations:

Example 1: My presentation will examine three main reasons why we should begin a technical communication degree program on this campus: (1) jobs are available in the area, (2) salaries are high, and (3) costs for starting the program are low in comparison to those for the more technical programs on campus.

Example 2: Tanya Ruskin richly deserves the award of employee of the year. Within the company she led the Business Department's successful effort to install the new accounts-receivable system. And outside the firm, she played a key role in organizing a volunteer team to help build a home sponsored by Habitat for Humanity.

The abstract structure suggested here gives listeners a clear sense of direction. They know what topic you will cover, why it is important, and what points will be included. Although listeners prefer this level of clarity, some speakers have occasional doubts about such an "up-front" structure. They think it's more effective to "tease" listeners at the beginning by only hinting at the subject, which will become clear later.

This kind of doubt is speaker centered, not audience centered. Remember that listeners lack familiarity with the material. That's why you're giving the speech to them, not the other way around! Thus the ambiguities they face should be resolved as quickly as possible with the PIP-centered abstract suggested here. The abstract simply provides a mental framework into which the audience can insert details covered in the body of the presentation.

BODY SECTION

As Figure 3–1 shows, the body of the speech is the longest, most detailed section. Here you provide descriptive details (for an informative speech), marshal your strongest arguments (for a persuasive speech), or supply your supporting material (for an occasional speech). The next chapter provides you with specific suggestions for informative, persuasive, and occasional speeches. Following are only some general recommendations that apply to the body of all oral presentations.

Body Guideline 1: Use Three to Five Main Sections

You may have observed that your memory works best when items are grouped into just a few categories. The same holds true for your listeners, who are searching for convenient "handles" to aid them in understanding your speech. If you give them just a few categories into which to place details, they are more likely to remember material from the body of the speech. Here are some examples of grouping:

- **Informative speech:** Five main skills you learned while completing an internship at the county courthouse
- **Persuasive speech:** Three problems that will be solved by adopting your proposal for a new media center at the college you attend
- **Occasional speech:** Four main contributions that support giving the "best student athlete" award to Morgan Smith, a graduating senior

In most cases, you already will have included the three to five main points in the "roadmap" statement at the end of the abstract. Therefore, the audience has been given an outline that is now reinforced by the structure of the body. The three to five points will, of course, appear in the same order in the body as they were presented in the abstract.

Body Guideline 2: Choose the Most Appropriate Pattern

Each speech requires that you choose an organizational pattern for the body that best suits the particular speech. The next chapter includes techniques related specifically to each of the three main speech types. Following are some general patterns that can be used in any speech:

- Question/answer
- Cause/effect
- Problem/solution
- Sequence of events
- Topics of importance
- Parts of objects

Spend some time thinking about the kind of pattern that would be most appealing to the audience. In an informative speech about your favorite job, the structure of the speech discussion could be based on (1) main job duties, (2) the sequence of tasks you confront during an average day, (3) the reasons why you chose this career, (4) lessons learned from the job, or (5) problems encountered during your time in the position. Your choice of a pattern would be determined by the purpose of the speech and needs of the audience.

Body Guideline 3: Use a Mini-ABC Format in Each Main Section

Just as the ABC format helps organize the entire speech, it also helps you reveal a sense of order in each of the main sections. Think of each section as a separate unit that, when assembled, produces a complete speech. Each of these mini-ABC units might do the following:

Abstract: Briefly state the main supporting question or statement.

Body: Give detailed support.

Conclusion: Summarize the point before moving on to the next one.

The main advantage of this "wheels within wheels" approach to organizing the body is that it provides listeners with a series of "mini-summaries," which are especially useful in more complex speeches.

Body Guideline 4: Use Many Transitional Devices

Certainly the use of mini-ABC formats helps maintain a sense of organization and keeps the audience attentive during your speech. There are also linguistic devices that can help with providing a smooth flow to your organizational plan. Here are a few transitional techniques, several of which will be covered in the chapter on delivery:

- Words that indicate sequence (first, second, third . . .)
- Words that indicate contrast (however, on the other hand)
- Pauses between major supporting points
- Distinct gestures at points of transition

Transitional devices help in several ways. First, they provide the auditory "glue" that makes the speech hang together as a coherent unit. Second, they present opportunities for you to recapture the attention of listeners who may have drifted off during the previous explanation.

Body Guideline 5: Follow Every Abstraction with Specific Examples

Good speeches always depend on effective use of vivid, concrete information to support abstract points. Examples, anecdotes, stories, analogies, and illustrations drive home your points better than any other device. Such details keep listeners engaged and help them recollect what you said. Here are some examples of abstract points, followed by specific support to back them up:

Abstraction: You enjoy gardening.

Concrete support: You include a story about how your parents bought you gardening tools and gave you a plot to garden when you were 10.

Abstraction: Fish oil prevents heart disease.

Concrete support: You refer to the Eskimo culture where salmon is a major part of the diet and where heart disease is rare.

Abstraction: Your college roommate had a strange sense of humor.

Concrete support: He once filled your car with turkey feathers before you were about to leave for a formal dance with your date.

As with writing, speeches are best received and remembered when they include a liberal application of vivid supporting examples.

Conclusion Section

Every speech conclusion has the same main objective: to summarize the speech and give the audience something to remember. Although there is no standard pattern of organization for accomplishing this objective, follow these general guidelines:

Conclusion Guideline 1: Restate the Main Three to Five Points

Repeating the main points satisfies the last part of the preacher's maxim mentioned earlier: "Then you tell 'em what you told 'em." Audience members need to be reminded of your main ideas because they probably have no written list in their hands. Of course, this summary can be reinforced by a graphical representation of the list, if you choose to include a visual at this point.

Conclusion Guideline 2: Indicate What Happens Next

"Action statements" are appropriate for many speeches, especially those that aim to persuade. Following are some action statements that would fit in all three types of speeches:

Action statement for informative speech: Now that you know more about the World Wildlife Fund, here's how to contact the local chapter if you'd like to help with a project in your community.

Action statement for persuasive speech: If you agree that more of our transportation tax should be spent on a rail system and less on highways, consider showing up at a hearing being held at the courthouse next Thursday.

Action statement for occasional speech: Given a lifetime of support that Gordon Rice provided the YMCA, I hope you will consider continuing his work by making a financial contribution to the organization.

Conclusion Guideline 3: Add a Personal Note

The very end of the speech gives you a moment to back out of the structured setting and finish with a personal comment, when appropriate. One example might be the kind of appeal to action mentioned in the previous guideline. Another might be a quick story, favorite quotation, or other device that reveals your own "humanity."

OUTLINES

The previous sections dealt mostly with the *product* of your effort to organize a speech—the ABC format and its use in a speech. This section instead deals mostly with the *process* by which you organize a speech—the outline. Two subsections follow. One provides a rationale for using outlines, and the other includes guidelines for using them during the process of preparing speeches.

REASONS FOR OUTLINES

Outlines have a dual purpose in your preparation for speeches. First, they remain the single best way for you to plan a presentation. Second, they may become the document you actually use for notes during an extemporaneous speech. This section examines three main reasons why you should develop an outline for every speech: organization, visualization, and review.

Organization

Outlines force you to grapple with matters of organization at a time when it is still easy for you to change the structure of the speech—that is, before you have begun writing the speech draft. As you add and delete ideas on the page, you are constantly thinking about the best way to respond to the needs of your audience.

Visualization

An outline shows you—visually—whether you have enough supporting information for particular parts of the speech. For example, if your outline includes only one subheading for a topic, you will notice this problem on the page and seek ways to address it—either by expanding the topic or deleting it. In either case it is the concrete visibility of the outline that prompts you to address the problem of a topic that is developed inadequately.

Review

Sometimes your speeches may need to be reviewed by another person or persons in your organization—for example, when you are speaking on a new product or service that needs to be presented with the utmost precision and care. Outlines speed up this review process. That is, it's much easier to make changes at the outline stage than later, after you have invested more time in the drafting stage. Reviewers who "sign off" on an outline are much less likely to request changes later.

Guidelines for Outlines

You might recall having to produce "tidy" outlines earlier in school. Each sheet probably had to include neatly arranged points sorted by Roman numerals, letters, and Arabic numbers. Such memories give outlines a bad name. The fact is that outlines are more useful as part of an organizational process than they are as a prettified "product." The following guidelines emphasize this process-oriented purpose.

Outline Guideline 1: Place Your Purpose Statement at the Top

Having the purpose of the speech at the top of the page grounds your outline before you even start. This purpose will guide your efforts as you strive to flesh out the main points.

Outline Guideline 2: Record Random Ideas Quickly

At this stage you do not have to rank your ideas. Just jot down all of the major and minor points that relate to your speech topic. This "nonlinear" process involves the free association of ideas that can only occur before you have applied structure to your thinking. Such an early creative step leaves you open to new ways to consider developing the topic.

Outline Guideline 3: Show Relationships

After you've gone through the previous step once or twice, you will have produced a sheet that contains most of the major and minor points to be included in the speech. Surveying the page, locate the major points that indicate the direction your speech will take and circle them. Then draw lines to connect these main supporting points to their supporting ones. Finally delete material you don't need. Figure 3–3 (Outline Process—Early Stage) shows the results of applying this process to a speech concerning problems and solutions with a company cafeteria.

Outline Guideline 4: Draft a Final Outline

Once related points are clustered, your messy outline can be cleaned up to make it useful to you as you write. You can use a traditional format with Roman

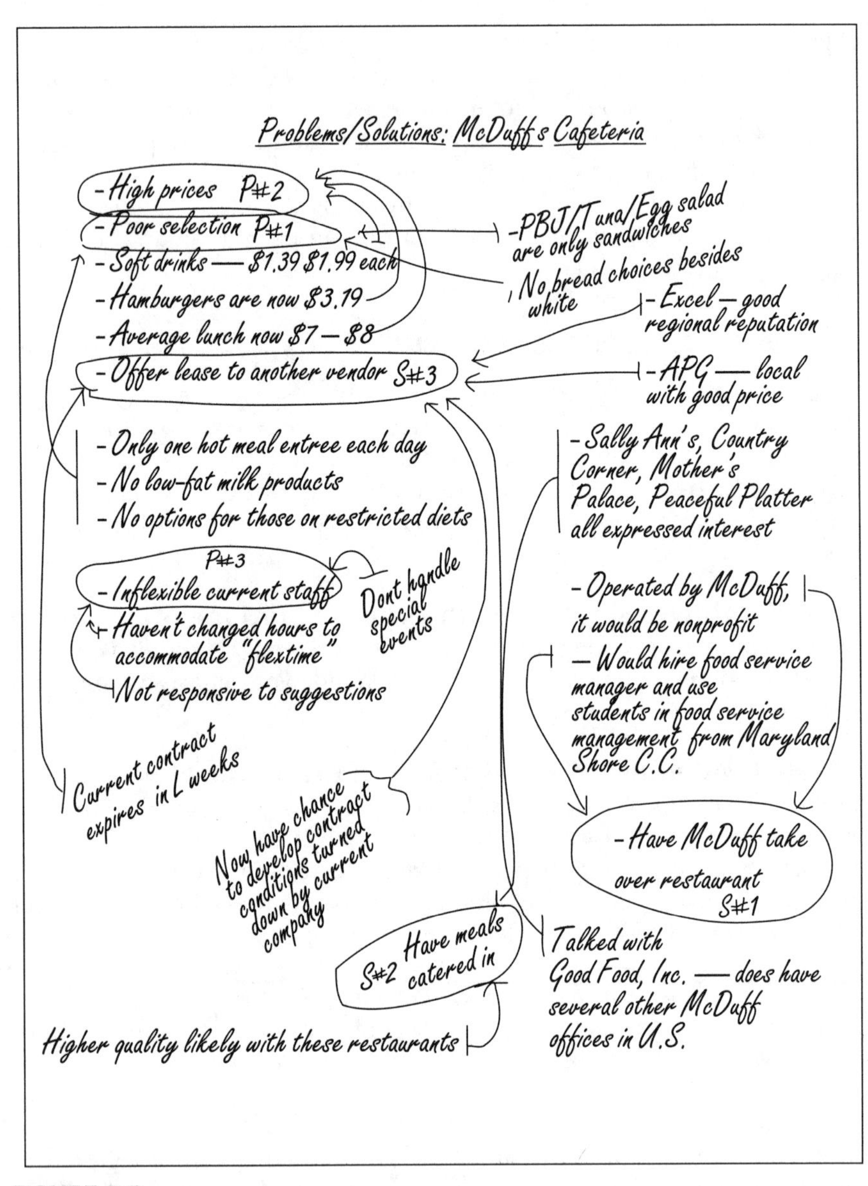

FIGURE 3–3
Outline process—early stage.

numerals as shown in Figure 3–4 (Outline Process—Later Stage), or you can simply use dashes and indenting to indicate outline levels. Whatever outline format you use, the final version should reflect the following features:

- **Depth:** The outline should include enough supporting information to develop the speech adequately.
- **Balance:** All main points should be supported by an adequate amount of detail. Note especially that when you subdivide a point, you will have two or more breakdowns—*any* object that has been divided will have at least two parts.
- **Parallelism:** Points in the same grouping should have the same grammatical form, so that it will be easier for you to make the transition later from outline to text. In fact, it's best to use either full sentences or phrases consistently throughout the entire outline. Phrases are preferable because they take up less space and don't lock you into particular wording too early in the process.

Outline Guideline 5: Plan Where to Use Graphics

The time to consider using graphics in your speech is early—at the outline stage. In this way, graphic communication will combine with text to become an integral part of the speech, not a tacked-on feature. In the late-stage outline in Figure 3–4, for example, you may see several possibilities for reinforcing main points with graphics. Here are a few:

- **Chart:** Showing the increase in cafeteria prices over the last three years
- **Table:** Contrasting prices for a few lunch items at the current cafeteria and at the four restaurants mentioned in Section V.B. of the outline
- **Map:** Showing the location of the four nearby restaurants
- **Chart:** Showing the relative costs for the contract with the vendors listed on Section VI.A. of the outline

Chapter 5 will give specific guidelines for using and developing graphics to support the text of your speeches.

CHAPTER SUMMARY

This chapter describes the ABC format, an organizational pattern that applies to all speeches. The ABC format places special emphasis on the beginning and ending sections of the speech. The abstract follows the PIP formula by presenting information related to purpose, importance, and plan. The body follows through on the promise of the abstract by developing three to five points in an order and pattern appropriate to the topic. Some suggested

PROBLEMS AND SOLUTIONS: CURRENT CAFETERIA IN BUILDING

I. Problem #1: Poor selection
  A. Only one hot meal entree each day
  B. Only three sandwiches—PBJ, egg salad, and tuna
  C. Only one bread—white
  D. No low-fat milk products (milk, yogurt, LF cheeses, etc.)
  E. No options for those with restricted diets
II. Problem #2: High prices
  A. Soft drinks from $1.39 to $1.99 each
  B. Hamburgers now $3.19
  C. Average lunch now $7–$8
III. Problem #3: Inflexible staff
  A. Unwilling to change hours to meet McDuff's flexible work schedule
  B. Have not acted on suggestions
  C. Not willing to cater special events in building
IV. Solution #1: End lease and make food service a McDuff department
  A. Hire food service manager
  B. Use students enrolled in food service management program at Maryland Shore Community College
  C. Operate as nonprofit operation—just cover expenses
V. Solution #2: Hire outside restaurant to cater meals into building
  A. Higher quality likely
  B. Initial interest by four nearby restaurants
    1. Sally Ann's
    2. Country Corner
    3. Mother's Palace
    4. Peaceful Platter
VI. Solution #3: Continue leasing space but change companies
  A. Initial interest by three vendors
    1. Excel—good regional reputation for quality
    2. APG—close by and local, with best price
    3. Good Food, Inc.—used by two other McDuff offices with good results
  B. Current contract over in two months
  C. Chance to develop contract not acceptable to current company

FIGURE 3–4
Outline process—later stage.

techniques are mini-ABC formats, transitions, and use of specifics after each abstraction. Finally, the conclusion restates the main points, indicates what happens next, and adds a personal touch. The speech process starts with the development of an outline that aids in organizing, visualizing, and reviewing the speech material. The best outlines result from a process that begins with random ideas on paper and ends with an organized hierarchy of points, leading you to the next step of writing the speech text.

EXERCISES

Exercise 1 can be completed by preparing either an oral or written report, depending on the preferences of your instructor. The other exercises require a written response.

1. **Speech Analysis—ABC Format**
   Choose the text of a speech from a source such as *Vital Speeches.* Analyze the degree to which the speech follows the ABC format described in this chapter. That is, describe how it meets or departs from the specific guidelines for using an abstract, a body, and a conclusion.
2. **Speech Outline—Informative Speech**
   Choose an informative speech topic about which you already have some information, such as your college major or a previous or current job. Construct a complete final draft of an outline. To show the process by which you developed the outline, submit both the "nonlinear" sheets and the final version.
3. **Speech Outline—Persuasive Speech**
   Choose a persuasive speech topic that relates to a current issue on your campus, at your place of employment, or in your community. Construct a complete final draft of an outline. To show the process by which you developed the outline, submit both the "nonlinear" sheets and the final version.
4. **Speech Outline—Occasional Speech**
   Choose as an occasional speech topic the introduction of someone you know reasonably well. Assume that your introductory remarks are being given to an audience that does not know the individual. Construct a complete final draft of an outline. To show the process by which you developed the outline, submit both the "nonlinear" sheets and the final version.

4 Text

This chapter describes some standard patterns of organization used to develop the body sections of informative, persuasive, and occasional speeches. Two points need to be made at the outset. First, some patterns more obviously apply to a particular speech type—for example, you will find the argumentative pattern most often in persuasive speeches and less often in informative speeches. Second, these same patterns are evident in written as well as spoken language. In fact, the title of this chapter is meant to draw attention to the fact that both speeches and documents use similar patterns in their discussion sections.

This book has emphasized that the best speeches are usually extemporaneous, or delivered from notes in an engaging manner that reflects the speaker's familiarity with the material. To reach the stage of writing good notes, most of us develop the speech in one of the following ways:

- Producing a detailed outline, which then leads to a set of notes to be used in the speech
- Writing out a complete draft of the text, which then leads to a set of notes to be used in the speech

This chapter is most useful if you plan to write out the text of your speech before generating notes, but it can also be used to create a detailed outline. In either case, the following patterns represent the most common ways to build the body section of the ABC format: definition, description, classification/division, comparison/contrast, and argument.

DEFINITION

Many speeches include terms that need to be defined for the audience. This section provides background information on definitions as well as guidelines for incorporating them into your speeches.

BACKGROUND ON DEFINITION

During your career, you often may use technical terms known only to those in your profession. As a civil engineer, for example, you would know that a "triaxial compression test" helps determine the strength of soil samples. As a computer professional, you would know the meaning of RAM (random-access memory), ROM (read-only memory), and LAN (local area network). When speaking to an audience that is unfamiliar with these fields, however, you may need to define technical terms.

Good definitions can support findings, conclusions, and recommendations throughout your speech. They also keep readers interested. Conversely, the most organized, well-written speech will fall on deaf ears if it includes terms that readers do not understand. For your listeners' sake, then, you need to be asking questions like these about definitions:

- How often should you use them?
- Where should they be placed?
- What format should they take?
- How much information is enough and how much is too much?

To answer these questions, the following section gives guidelines for using definitions in your oral presentations.

Guidelines for Definition

Once you know definitions are needed in your speech, you must decide on their format and location. Again, consider your audience. How much information do listeners need? Where is this information best placed within the speech? Following are five guidelines for writing good definitions.

Definition Guideline 1: Keep It Simple

On rare occasions, the sole purpose of a speech is to define one or more terms. Most of the time, however, a definition just clarifies a term in a speech that has a larger purpose. Your definition should be as simple and unobtrusive as possible. Present only the level of detail needed by the listener.

For example, in speaking to clients on your land survey of their industrial site, you might briefly define a transit as "the instrument used by land surveyors to measure horizontal and vertical angles." The main purpose of the speech is to present property lines and total acreage, not to give a lesson in surveying, so this sentence definition is adequate. Choose from these three main formats (listed from least to most complex) in deciding the form and length of definitions:

- **Informal definition:** A word or brief phrase that gives only a synonym or other minimal information about the term
- **Formal definition:** A full sentence that distinguishes the term from other similar terms and that includes these three parts: the term itself, a class to which the term belongs, and distinguishing features of the term
- **Expanded definition:** A lengthy explanation that begins with a formal definition and is developed into several paragraphs or more

Guidelines 2–4 show you when to use these three options in your speech.

Definition Guideline 2: Use Informal Definitions for Simple Terms Most Listeners Understand

Informal definitions appear immediately after the terms being defined, often as one-word synonyms. They give just enough information to keep the listener minimally informed about the topic. As such, they are best used with simple terms that can be defined adequately without much detail.

Here is a situation in which an informal definition would apply. Your firm has been hired to examine a possible shopping-mall site. The buyers, a group of physicians, want a presentation that includes a list of previous owners and an opinion about the suitability of the site. As a legal assistant at your firm, you must assemble a list of owners as your part of the group project. You want your speech to agree with court records, so you decide to include real-estate jargon such as "grantor" and "grantee." For your nontechnical listeners, you include definitions like the following:

> All **grantors** (persons from whom the property was obtained) and **grantees** (persons who purchased the property) are listed on the following chart, by year of ownership.

This same speech has a section describing creosote pollution found at the site. The chemist presenting the contamination section also uses an informal definition for the benefit of the listener:

> At the southwest corner of the mall site, we found sixteen barrels of **creosote** (a coal tar derivative) buried under about 3 feet of sand.

The listeners do not need a fancy chemical explanation of creosote. They only need enough information to keep them from getting lost in the terminology. Informal definitions perform this task nicely.

Definition Guideline 3: Use Formal Definitions for More Complex Terms

A formal definition appears in the form of a sentence that lists (1) the **term** to be defined, (2) the **class** to which it belongs, and (3) the **features** that distinguish the term from others in the same class. Use it when your audience needs more background than an informal definition provides. Formal definitions define in two stages:

- First, they place the term into a *class* (group) of similar items.
- Second, they list *features* (characteristics) of the term that separate it from all others in that same class.

In the list of sample definitions that follows, note that some terms are tangible (like "pumper") and others are intangible (like "arrest"). Yet all can be defined by first choosing a class and then selecting features that distinguish the term from others in the same class.

Term	*Class*	*Features*
An *arrest* is	restraint of persons	that deprives them freedom of movement and binds them to the will and control of the arresting officer.
A *financial statement* is	a historical report about a business	that is prepared by an accountant to provide information useful in making economic decisions, particularly for owners and creditors.
A *triaxial compression test* is	a soils lab test	that determines the amount of force needed to cause a shear failure in a soil sample.
A *pumper* is	a fire-fighting apparatus	used to provide adequate pressure to propel streams of water toward a fire.

This list demonstrates three important points about formal definitions. First, the definition itself must not contain terms that are confusing to your readers. The definition of "triaxial compression test," for example, assumes listeners will understand the term *shear failure* that is used to describe features. If this assumption were incorrect, then the term would need to be defined. Second, formal definitions may be so long that they create a major distraction in the speech. Third, the class must be narrow enough so that you will not have to list too many distinguishing features.

Definition Guideline 4: Use Expanded Definitions for Supporting Information

Sometimes a parenthetical phrase or formal sentence definition is not enough. If readers need more information, use an expanded definition with this three-part structure:

- **An overview at the beginning**—Including a formal sentence definition and a description of the ways you will expand the definition
- **Supporting information in the middle**—Using listings on graphics as helpful format devices for the audience
- **Brief closing remarks at the end**—Reminding the reader of the definition's relevance to the whole speech

Here are seven ways to expand a definition, along with brief examples:

1. **Background and/or history of term:** Expand the definition of "triaxial compression test" by giving a dictionary definition of "triaxial" and a brief history of the origin of the test.

2. **Applications:** Expand the definition of "financial statement" to include a description of the use of such a statement by a company about to purchase controlling interest in another company.
3. **List of parts:** Expand the definition of "pumper" by listing the parts of the device, such as the compressor, the hose compartment, and the water tank.
4. **Graphics:** Expand the description of the triaxial compression test with an illustration showing the laboratory test apparatus.
5. **Comparison/contrast:** Expand the definition of a term like "management by objectives" (a technique for motivating and assessing the performance of employees) by pointing out similarities and differences between it and other management techniques.
6. **Basic principle:** Expand the definition of "ohm" (a unit of electrical resistance equal to that of a conductor in which a current of one ampere is produced by a potential of 1 volt across its terminals) by explaining the principle of Ohm's Law (that for any circuit the electric current is directly proportional to the voltage and inversely proportional to the resistance).
7. **Illustration:** Expand the definition of CAD/CAM (computer-aided design/computer-aided manufacturing techniques to automate the design and manufacture of products) by giving examples of how CAD/CAM is changing methods of manufacturing many items, from blue jeans to airplanes.

Obviously, long definitions might seem unwieldy within an oral presentation. For this reason, you need to think long and hard before you spend time on them. Most speakers prefer to give short definitions. Then they can always provide a handout of definitions after the speech, for later reference by the audience.

Definition Guideline 5: Choose the Right Location for your Definition

Short definitions are likely to be in the body of the speech; long ones are often relegated to technical handouts, as just noted. However, length is not the main consideration. Think first about the *importance* of the definition to your audience. If you know that decision makers will need the definition, then place it in the speech—even if it is fairly lengthy. If the definition only provides supplementary information, then it can be eliminated or placed in a handout. You have these five choices for locating a definition:

1. **In the same sentence as the term,** as with an informal, parenthetical definition
2. **In a separate sentence,** as with a formal sentence definition occurring right after a term is mentioned
3. **In a footnote to a speech produced as text,** as with a formal or expanded definition listed at the bottom of the page on which the term is first mentioned

4. **In a glossary at the beginning or end of a speech produced as text,** along with all other terms needing definition in that document
5. **In an appendix at the end of a speech produced as text,** as with an expanded definition that would otherwise clutter the text of the speech

Clearly, only options 1 and 2 are usually heard by the audience during the speech. The others provide reference material for later consideration.

DESCRIPTION

Listeners prefer presentations that come alive with physical details that can be visualized. In fact, many people consider themselves extremely "visual" in orientation. They expect language that helps them picture ideas and things. This section gives additional background information on descriptions and offers guidelines for placing them in your speech.

BACKGROUND ON DESCRIPTION

Technical descriptions, like expanded definitions, require that you pay special attention to *details.* In fact, you can consider a description to be a special type of definition that focuses on parts, functions, or other features. It emphasizes *physical* details. Descriptions can appear in any part of a speech, from the introduction to the conclusion. As a rule, however, detailed descriptions tend to be placed in body sections as supporting information.

Following is an example of a description that might appear in the body of a speech. Assume your environmental firm has been hired to recommend alternative locations for a new swimming and surfing park in southern California. Delivered to a county commission (five laypersons who will make the final decision), your speech will give three possible locations and your reasons for selecting them as options. The body of the speech will include the following descriptive information about each site: surface features, current structures on the sites, types of soils gathered from the surface, water quality, and aesthetic features. In other words, the descriptions in the body of the speech will support your overall purpose.

GUIDELINES FOR DESCRIPTION

Now that you know how descriptions can be used in speeches, following are guidelines for using accurate, detailed descriptions. Follow them carefully as you prepare assignments in class and on the job.

Description Guideline 1: Remember Your Listeners' Needs

The level of detail in a technical description depends on the purpose it serves. Give the audience precisely what it needs but no more. In the previous speech about park locations, the commissioners do not want a detailed description of soil samples taken from borings. That level of detail will be reserved for a few sites selected later for further study. Instead, they want only surface descriptions. Always know just how much detail will get the job done.

Description Guideline 2: Be Accurate and Objective

More than anything else, listeners expect accuracy in descriptions. Pay close attention to details. (As noted previously, the *degree* of detail in a description depends on the *purpose* of the speech.) Along with accuracy should come objectivity. This term is more difficult to pin down, however. Some speakers assume that an objective description leaves out all opinion. Such is not the case. Instead, an objective description may include opinions that have the following features:

- They are based on your professional background.
- They can be supported by details from the site or object being described.

Description Guideline 3: Choose an Overall Organization Plan

Like other patterns discussed in this chapter, technical descriptions usually make up only parts of speeches. Yet they should have an organization plan that permits them to be delivered as self-contained, stand-alone sections. Following are three common ways to describe physical objects and events. In all three cases, a description should move from general to specific. That is, you begin with a view about the entire object or event. Then in the rest of the description, you focus on specifics.

- **Description of the parts:** For many physical objects, you will simply organize the description by moving from part to part.
- **Description of the functions:** Often the most appropriate overall plan relies on how things work, not on how they look. A function-oriented description would include only a brief description of the parts.
- **Description of the sequence:** If your description involves events, as in a police officer's speech about an accident investigation, you can organize ideas around the major actions that occurred, in their correct sequence. As with any list, it is best to place a series of many activities into just a few groups. Four groups of five events each is much easier for listeners to comprehend than a single list of twenty events.

Description Guideline 4: Use "Helpers" Like Graphics and Analogies

The words of a technical description need to come alive. Because your listeners may be unfamiliar with the item, you should search for ways to connect with their experience and with their senses. Two effective tools are graphics and analogies.

Graphics respond to the desire of most listeners to see pictures along with hearing words. As the listeners listen to your part-by-part or functional breakdown of a mechanism, they can refer to your graphic aid for assistance. An illustration helps you too, of course, in that you need not be as detailed in describing locations and dimensions of parts when you know the audience has easy access to a visual.

Analogies, like illustrations, give listeners a convenient handle for understanding your description. Put simply, an analogy allows you to describe something unknown or uncommon in terms of something that is known or more common. A brief analogy can sometimes save you hundreds of words of technical description. This paragraph description contains three analogies:

> KB-Tech, Inc., is equipped to help clean up oil spills with its patented product, SeaClean. This highly absorbent chemical is spread over the entire spill by means of a helicopter, which makes passes over the spill much like a lawn mower would cover the complete surface area of a lawn. When the chemical contacts the oil, it acts like sawdust coming in contact with oil on a garage floor. That is, the oil is immediately absorbed into the chemical and physically transformed into a product that is easily collected. Then our nearby ship can collect the product, using a machine that operates much like a vacuum cleaner. This machine sucks the SeaClean (now full of oil) off the surface of the water and into a sealed container in the ship's hold.

Description Guideline 5: Give your Description the "Visualizing Test"

After completing a description, test its effectiveness by reading it to someone who is unfamiliar with the material—someone with about the same level of knowledge as your intended audience. If this person can draw a rough sketch of the object or events while listening to your description, then you have done a good job. If not, ask your listener for suggestions to improve the description. If you are too close to the subject yourself, sometimes an outside point of view will help refine your technical description.

CLASSIFICATION/DIVISION

In oral presentations, you often perform these related tasks: (1) grouping lists of items into categories, a process called "classification," or (2) separating an individual item into its parts, a process called "division." In practice, the patterns usually work together. For the purposes of clarity, however, here the guidelines are presented separately, preceded by background information on the entire pattern.

Background on Classification/Division

Classification helps you (and, of course, your listener) make sense out of diverse but related items. For example, the process of outlining a speech requires the use of classification on your part. As noted in chapter 3, outlining forces you to apply both classification and division to organize diverse information into manageable "chunks."

Here's an example of the classification/division activity (mostly classification) applied to a speech setting. The human resources manager of a manufacturing firm wants to change the procedure whereby employees can register grievances. She wants a committee to be appointed to hear grievances of all kinds, and she plans to give a speech to the board of directors proposing this change. As part of her speech, she will present a summary of the types of grievances filed since the company was founded in 1954. Her research has uncovered 116 separate grievances for which there is paperwork. Her speech to the board will classify them into four categories: (1) firing or discharge from the firm; (2) use of seniority in layoffs, promotions, or transfers; (3) yearly performance evaluations; and (4) the issue of required overtime in some offices. Her speech will employ these classifications to support the need for a broad range of experience represented on the committee.

Guidelines for Classification/Division

The guidelines presented here provide a three-step procedure for classifying any group of related items. Use them as you prepare to construct a classification scheme that will be included in your oral presentations.

Classification Guideline 1: Find a Common Basis

Classification requires that you establish your groupings on one main basis. This basis can relate to size, function, purpose, or any other factor that serves to produce logical groupings. For example, the basis for the grievance case mentioned earlier is the *purpose* for the grievance. That is, the grievances can be grouped under four main reasons why the employees made formal complaints to the company.

Classification Guideline 2: Limit the Number of Groups

Listeners prefer groupings of three to seven items. The general rule is this: The fewer the better. This principle of organization, as well as the appropriateness of the basis, should guide your use of classification. Strive to select a basis that will result in a limited number of groupings.

Classification Guideline 3: Classify Each Item Carefully

The final step is to place each item in its appropriate classification. If you have chosen classifications carefully, this step is no problem. For example, assume that as a firm's finance expert, you want to identify sources for funding new projects. Therefore, you need to collect the names of all commercial banks started in Pennsylvania last year. A reference librarian gives you the names of forty-two banks. To impose some order, you decide to classify them by charter. The resulting groups are as follows:

- National banks (those chartered in all states)
- Non-Federal Reserve state banks (exclusively Pennsylvania banks that are not members of the Federal Reserve System)
- Federal Reserve state banks (exclusively Pennsylvania banks that are members of the Federal Reserve System)

With three such classifications that do not overlap, you can put each of the forty-two banks in one of the three classifications.

Guidelines for Division

Division begins with an entire item that must be *broken down* or *partitioned* into its parts, whereas classification begins with a series of items that must be *grouped* into related categories. Division is especially useful when you need to explain a complicated piece of equipment to an audience that is unfamiliar with it. Follow these three guidelines for applying this pattern:

Division Guideline 1: Choose the Right Basis for Dividing

Like classification, division means you must find a logical reason for establishing groups or parts. Assume, for example, that you are planning to teach a training seminar in project management. In dividing this four-day training seminar into appropriate segments, it seems clear to you that each day should cover one of these crucial parts of managing projects: meeting budgets, scheduling staff, completing written reports, and seeking follow-up work from the client. In this case, the principle of division seems easy to employ.

Yet other cases present you with choices. If, for example, you were planning a training presentation on report writing, you could divide it in these three ways, among others: (1) by *purpose of the report* (for example, progress, trip, recommendation), (2) by *parts of the writing process* (for example, brainstorming, outlining, drafting, revising), or (3) by *report format* (for example, letter report, informal report, formal report). Here you would have to choose the basis most appropriate for your purpose and audience.

Division Guideline 2: Subdivide Parts When Necessary

As with classification, the division pattern of organization can suffer from the "laundry list" syndrome. Specifically, any particular level of groupings should have from three to seven partitions—the number that most listeners find they can absorb. When you use more than that number, consider reorganizing information or subdividing it.

Assume that you want to give a presentation to your staff on the contents of a short manual on writing formal proposals. Your first effort to divide the topic results in nine segments: cover page, letter of transmittal, table of contents, executive summary, introduction, discussion, conclusions, recommendations, and appendices. You would prefer fewer groupings, so you then establish three main divisions: front matter, discussion, and back matter, with breakdowns of each. In other words, your effort to partition was guided by every listener's preference for a limited number of groupings.

Division Guideline 3: Describe Each Part with Care

This last step may seem obvious. Make sure to give equal treatment to each part of the item or process you have partitioned. Listeners expect this sort of parallelism, just as they prefer the limited number of parts mentioned in Guideline 2.

COMPARISON/CONTRAST

Many speeches require that you show similarities or differences between ideas or objects. (For our purposes, the word *comparison* emphasizes similarities, whereas the word *contrast* emphasizes differences.) In the real world of career speaking, this technique applies especially to situations wherein listeners are making buying decisions. This section presents background information and related guidelines for the pattern.

Background on Comparison/Contrast

You probably remember being asked to use this pattern of organization in writing and speaking assignments early in school. It is a common way to encourage individuals to think critically about choosing among options. Assume you have had the good fortune to travel to Jamaica three times for your company. Each time you stayed in a different hotel in Montego Bay. Now your firm has decided to hold its quarterly meeting there. Before selecting a hotel, your supervisor asks you to speak to the arrangements committee for about 20 minutes on the main features of the three hotels. Presumably, the committee will decide on a location on the basis of the information you present.

GUIDELINES FOR COMPARISON/CONTRAST

This pattern of organization undergirds many informative, persuasive, and occasional speeches. Guidelines follow for using the comparison/contrast pattern.

Comparison/Contrast Guideline 1: Remember Your Purpose

When using comparison/contrast in on-the-job speeches, your purpose usually falls into one of the following categories:

1. **Objective:** Essentially an unbiased presentation of features wherein you have no real "axe to grind"
2. **Persuasive:** An approach wherein you compare features in such a way as to recommend a preference

You must constantly remember your main purpose. You will either provide raw data that someone else will use to make decisions *or* you will urge someone toward your preference. In either case, the comparison must show fairness in dealing with all alternatives. Only in this way can you establish credibility in the eyes of the audience.

Comparison/Contrast Guideline 2: Establish Clear Criteria and Use Them Consistently

In any technical comparison, you must set clear standards of comparison and then apply them uniformly. Otherwise, your listener will not understand the evaluation or accept your recommendation (if there is one).

For example, assume you are a field supervisor who must give a speech recommending the purchase of a bulldozer for construction sites. You have been asked to recommend just *one* of these models: Cannon-D, Foley-G, or Koso-L. After background reading and field tests, you decide upon three main criteria for your comparison: (1) pushing capacity, (2) purchase details, and (3) dependability. These three factors, in your view, are most relevant to the company's needs. Having made this decision about criteria, you then must discuss all three criteria with regard to *each* of the three bulldozers. Only in this way can the audience get the data needed for an informed decision.

Comparison/Contrast Guideline 3: Choose the "Whole-by-Whole" Approach for Short Comparison/Contrasts

This strategy requires that you discuss one item in full, then another item in full, and so on. Using the bulldozer example, you might first discuss all features of the Cannon, then all features of the Foley, and finally all features of the Koso. Such a strategy works best if individual descriptions are quite short so readers can remember points made about the Cannon bulldozer as they proceed to hear sections on the Foley and then the Koso machines.

Keep these two points in mind if you select the whole-by-whole approach:

- Discuss subpoints in the same order—that is, if you start the Cannon description with information about dependability, begin the Foley and Koso discussions in the same way.
- If you are making a recommendation, move from least important to most important, or vice versa, depending on what approach will be most effective with your audience. Busy listeners usually prefer that you start with the recommended item, followed by the others in descending order of importance.

Comparison/Contrast Guideline 4: Choose the "Part-by-Part" Approach for Long Comparison/Contrasts

Longer comparison/contrasts require listeners to remember much information. Therefore, they usually prefer that you organize the comparison around major criteria, *not* around the whole items. Using the bulldozer example, your speech would follow this outline:

I. Pushing Capacity
 A. Cannon
 B. Foley
 C. Koso

II. Purchase Details
 A. Cannon
 B. Foley
 C. Koso

III. Dependability
 A. Cannon
 B. Foley
 C. Koso

Note that the bulldozers are discussed in the same order within each major section. As with the whole-by-whole approach, the order of the items can go either from most important to least important or vice versa—depending on what strategy you believe would be most effective with your audience.

Comparison/Contrast Guideline 5: Use Illustrations

Comparison/contrasts of all kinds benefit from accompanying graphics. In particular, tables are an effective way to present comparative data. For example, your speech on the bulldozers might present some pushing-capacity data you found in company brochures.

ARGUMENT

Good argument forms the basis for many speeches. This section provides you with background information that includes a broad-based definition of argument. The guidelines for using argument follow.

BACKGROUND ON ARGUMENT

Some people have the mistaken impression that only speeches dealing with recommendations and proposals argue their case to the audience, and that all other speeches should be objective rather than argumentative. The fact is, almost every time you speak you are "arguing" your point. This book uses the following definition for argument:

> **Argument:** The strategies you use in presenting evidence to support your point *and* to support your professional credibility, while still keeping the goodwill of the audience.

Thus even the most noncontroversial speech, like a trip report, involves argument in the sense defined here. That is, a trip report would present information to support the fact that you accomplished certain objectives on a business trip. It also would show the audience that you worked hard to accomplish your objectives. In fact, you hope that every speech you write becomes an "argument" for your own conscientiousness as a professional.

Following is an example of how argument might be used in speaking. Assume that many of the workers in your large office have experienced regular back pain ever since the company purchased some expensive new office chairs six months ago. Unfortunately, your furniture was ordered by the corporate office, so you must go through channels to seek a solution. As branch manager, you've mentioned the problem to your boss, a vice president at the corporate office. You will be visiting corporate in two weeks for a meeting so he has asked you to make a short presentation on the problem to members of the executive staff. Any change would involve considerable expense they would have to approve. You don't want to convey a complaining attitude in this presentation. Instead, you wish to document thoroughly and objectively the problems associated with the new chairs.

GUIDELINES FOR ARGUMENT

Speeches focused on argument have a long tradition, from the rhetoric of ancient times to the political debate of today. This section describes six guidelines for argument that apply to on-the-job oral presentations.

Argument Guideline 1: Use Evidence Correctly

Whether presenting a speech to a customer or your company's board of directors, you often move from specific evidence toward a general conclusion supported by evidence. Called "inductive reasoning," this approach to argument requires that you follow some accepted guidelines. Here are three:

- Use points the reader can grasp
- Use points that are a representative sample
- Have enough points to justify your conclusion

Argument Guideline 2: Choose the Most Convincing Order for Points

Chapter 3 points out that listeners pay most attention to beginnings and endings. Thus, you should include your stronger points at the beginning and ending of an argumentative sequence, with weaker points in the middle. The strongest point usually appears:

- **Last**—in arguments that are fairly short
- **First**—in arguments that are fairly long

You can expect the audience to stay attentive in a short passage, as you lead up to your strongest point. For long passages, however, the audience may become less attentive. For this reason, place the most important evidence at the beginning of longer arguments, where the audience will hear it before they get distracted.

Argument Guideline 3: Be Logical

Some arguments rest on a logical structure called a *syllogism*. When it occurs formally (which is rare), a syllogism may look something like this:

- **Major premise:** All chairs with straight backs are unacceptable for prolonged office work.
- **Minor premise:** The chairs at the Boston office have straight backs.
- **Conclusion:** The chairs at the Boston office are unacceptable for prolonged office work.

In fact, syllogisms usually appear in an implied, less-structured fashion than the sample shown here.

Argument Guideline 4: Use Only Appropriate Authorities

In the technical world, experts often supply evidence that will win the day in both written and spoken communication. Be certain that the experts you have chosen have credentials that the audience will accept.

Argument Guideline 5: Avoid Argumentative Fallacies

Guidelines 1–4 mention a few ways you can err in arguing your point. Listed next are some other fallacies that can damage your argument. Accompanying each fallacy is an example from the case study mentioned earlier about chair problems:

- **Ad hominem** (Latin: "to the man"): Arguing against a person rather than discussing the issue. (Example: Suggesting that the chairs should be replaced because the purchasing agent at the Baltimore office, who ordered the chairs, is surly and incompetent.)
- **Circular reasoning:** Failing to give any reason for why something is or is not true, other than stating that it is. (Example: Proposing that the straight-backed chairs should be replaced simply because they are straight-backed chairs.)
- **Either/or fallacy:** Stating that only two alternatives exist—yours and another one that is much worse—when in fact there are other options. (Example: Claiming that if the office does not purchase new chairs next week, fifteen employees will quit.)
- **False analogy:** Suggesting that one thing should be true because of its similarity to something else, when in fact the two items are not enough alike to justify the analogy. (Example: Suggesting that the chairs in the Boston office should be replaced because the Atlanta office has much more expensive chairs.)
- **Hasty generalization:** Forming a generalization without adequate supporting evidence. (Example: Proposing that all Jones office furniture will be unacceptable, simply because the current chairs carry the Jones trademark.)
- **Non sequitur** (Latin: "it does not follow"): Making a statement that does not follow logically from what came before it. (Example: Claiming that if the office chairs are not replaced soon, productivity will decrease so much that the future of the office will be in jeopardy—even though there is no evidence that the chairs would have this sort of dramatic effect on business.)
- **Post hoc ergo propter hoc** (Latin: "after this, therefore because of this"): Claiming a cause–effect relationship between two events simply because one occurred before the other. (Example: Associating the departure of an excellent secretary with the purchase of the poorly designed chairs a week earlier—even though no evidence supports the connection.)

Argument Guideline 6: Refute Opposing Arguments

Often you will be faced with arguments that oppose your views. Perhaps your proposition will cost a great deal of money, take considerable time to complete, or risk alienating employees with a sudden change of procedure. Effective argument requires that you strive to do the following:

- Identify the strongest opposing points
- Counter these points with effective refutation

In the process, be certain to maintain an even, unemotional tone that does not irritate those with opposing views. Your goal is to convince with facts and analysis. For example, you might choose to counter a point about the significant short-term cost of your proposal with a response that emphasizes the long-term benefits to your organization.

CHAPTER SUMMARY

As noted in chapter 3, most speeches reflect an overall ABC pattern that includes an abstract (introduction), a body (discussion), and a conclusion (wrap-up). The discussion sections of informative, persuasive, and occasional speeches often use traditional patterns of organization to build support for their points. This chapter provides background information and guidelines for using the following patterns in your oral presentations: definition, description, classification/division, comparison/contrast, and argument.

EXERCISES

All of the exercises that follow can be completed as individual or group assignments, depending on the choice of your instructor.

1. **Speech That Uses Definition**
   Conduct some basic research on a career field of interest to you. Then prepare a speech that introduces your audience to the field. Your speech should include definitions of at least three terms that would be unfamiliar to a general audience.
2. **Speech That Uses Description**
   Interview a friend or colleague about the job that person holds. Be certain it is a job that you yourself do *not* have. On the basis of the information you have collected, prepare a speech that describes the person's job—major responsibilities, reporting relationships, educational preparation, experience required, etc. Assume that your audience is an outside consulting firm that is evaluating the company's compensation sys-

tem. The goal of the evaluation is to achieve greater correlation between job responsibilities and pay grades, so the group needs accurate information about the position.

3. **Speech That Uses Classification/Division**
For this speech you will need a catalog from a college or university. First, select a number of courses that have not yet been classified except perhaps by department. Using course descriptions in the catalog or other information you can find, classify the courses using an appropriate basis (for example, the course level, topic, purpose in the department, prerequisites, etc.). Also, partition one of the courses by using information gathered from a course syllabus or from an interview of someone familiar with the course. Now present the information in a speech, as if you were a university employee addressing a group of people who are already employed but who are interested in further education in the areas covered. Your speech will help these people decide whether to enroll.
4. **Speech That Uses Comparison/Contrast**
Select a product category approved by your instructor. Then choose three specific types of brand-name products that fit within the category and for which you can find data. Prepare either a whole-by-whole or a part-by-part comparison/contrast speech. Assume that you are a research assistant for a large multinational firm. You are giving your presentation to a group of purchasing managers from around the world who work at various offices of the firm. The purpose of the speech is to give them the information they need to choose one of the products over the others.
5. **Speech That Uses Argument**
Prepare a speech that makes an effective argument for one of the following cases or for another proposition:

    - Purchasing a particular product for your company
    - Selecting a particular career field
    - Choosing a particular country or city as a site for living or working
    - Starting or ending a particular extracurricular activity at your school
    - Adopting a particular approach to physical fitness
    - Supporting a particular candidate for local election
    - Attending a particular college or university

5

Graphics

General Guidelines for Speech Graphics
- Preparing Graphics
- Using Graphics

Specific Guidelines for Eight Graphics
- Tables
- Pie Charts
- Bar Charts
- Line Charts
- Schedule Charts
- Flowcharts
- Organization Charts
- Technical Drawings

Misuse of Graphics
- Description of the Graphics Problem
- Examples of Distorted Graphics
- Special Issues Related to PowerPoint

Chapter Summary

Exercises

More than ever before, listeners expect good graphics during oral presentations. Graphics can help transform the words of your speech into true communication with the audience. The key word here is "help." Too many speakers have decided that fancy graphics can substitute for features that always have been, and always will be, the hallmarks of an effective presentation—such as clear organization, enthusiastic delivery, and solid content. Fancy graphics cannot disguise an otherwise mediocre presentation, but they can enhance one that is already good. The purpose of this chapter is to show you how graphics can enhance your presentation.

The first section of the chapter provides six general guidelines for preparing and using graphics within your presentation. The second section gives specific rules for producing eight common types of graphics often used in oral and in written reports: tables, pie charts, bar charts, line charts, schedule charts, flowcharts, organization charts, and technical drawings. These eight types are among the most common from which to choose for speeches. The last part of the chapter shows you how to avoid misuse of graphics, with special emphasis on problems in the use of electronic graphics such as PowerPoint.

Graphics create a powerful effect on your audience if used skillfully. Let's begin by reviewing some basic guidelines that will help you use them well.

GENERAL GUIDELINES FOR SPEECH GRAPHICS

The best all-around suggestion for graphics is to view them as an integral part of the speech. This level of coordination can only be achieved if you plan graphics in the same way you plan your text—long before the speech is given. Think of text and pictures as a "team." The suggestions in this section apply to all graphics used in speeches, from the simple to the complex. The first set of guidelines can be implemented while you are preparing the speech, whereas the second set is implemented during delivery.

PREPARING GRAPHICS

As noted in the chapter 3 discussion of outlines, you should plan where to use visuals during the process of writing your speech outline. Actually, consideration of graphics may occur even before this stage during your first musings about the speech. Following are six suggestions to follow early in the process.

Graphics Planning Guideline 1: Discover Listener Preferences

Some speakers prefer simple graphics, such as a flip chart. Others like sophisticated equipment, such as video projectors connected to laptop computers or multiple slide projectors. Listeners usually indicate their preferences when you ask them. They may also indicate preferences as to the type of graphic design—for example, tables, bar charts, or drawings. Contact key members of the audience ahead of time and make some inquiries about preferences. Then follow up with the event coordinator by doing the following:

- Ask for information about the room in which you will be speaking and any peculiarities related to setting up equipment.
- Request a setting that allows you to make best use of the graphics preferences of your audience.
- Choose graphics that best fit the constraints of the room if your audience has no known preferences (details about lighting, wall space, and chair configuration can influence your selection).

Graphics Planning Guideline 2: Think about Graphics Early

Graphics prepared as an afterthought usually look "tacked on." Plan them while you prepare the text so that the presentation will seem fluid. This guideline holds true especially if you rely upon specialists to prepare your visuals. Professionals need some lead time to do their best work. Also, they can often provide helpful insights about what visuals will enhance your presentation—if you consult them early enough and if you make them part of your presentation team. Never put yourself in the position of having to apologize for the quality of your graphic material.

Graphics Planning Guideline 3: Keep the Message Simple

Listeners are suspicious of overly slick visual effects. Many people prefer the simplicity of overhead transparencies with large, clear wording, such as those used in the speech example in the Appendix. If you use sophisticated technology, make sure it fits your context and audience.

Graphic Planning Guideline 4: Make Wording Brief and Visible

Some basic design guidelines apply whether you are using posters, overhead transparencies, or computer-aided graphics such as PowerPoint:.

- Use few words, emphasizing just one idea on each frame.
- Use a lot of white space, perhaps as much as 60–70 percent per frame.
- Use "landscape" format more often than "portrait," especially because it is the preferred default setting for much presentation software.
- Use sans serif print (the letters without "tails") because it is generally considered more readable from a distance.
- Use large enough print (14 point to 48 point) to be visible around the room where you are presenting.
- Use upper- and lowercase—not full caps—because it is considered more legible from a distance.

Graphics Planning Guideline 5: Use Colors Carefully

Colors can add flair to visuals. Follow these simple guidelines to make colors work for you:

- Have a good reason for using color (such as the need to highlight three different bars on a graph with three distinct colors).
- Use only dark, easily seen colors, and be sure that a color contrasts with its background (for example, yellow on white does not work well).
- Use no more than three or four colors in each graphic (to avoid a confused effect).
- For variety, consider using white on a black or dark green background.

Graphics Planning Guideline 6: Practice Using Your Graphics

Include every graphic you plan to use in your practice sessions. This is a good reason to prepare graphics as you prepare text, rather than as an afterthought. Running through a final practice without graphics would be like doing a dress rehearsal for a play without costumes and props—you would be leaving out parts that require the greatest degree of timing and orchestration.

Using Graphics

Once you have prepared and practiced with the graphics, you're ready for the real thing. Following are some suggestions for handling graphics during the presentation itself.

Graphics Use Guideline 1: Leave Graphics Up the Right Amount of Time

Because graphics reinforce text, they should be shown only while you address the related point in the text of the speech. For example, reveal a graph just as you are saying, "As you can see from the graph, the projected revenue reaches a peak in 2005." Then pause and leave the graph up a bit longer for the audience to absorb your message. Remember these points:

- A graphic outlives its usefulness when it remains in sight after you have moved on to another topic. Listeners will continue to study it and ignore what you are now saying.
- If you use a graphic once and plan to return to it, take it down after its first use and show it again later.

Graphics Use Guideline 2: Avoid Handouts

Because keeping listeners attentive is crucial, it's best not to use handouts while you are speaking. Otherwise the attention of the listeners will move away from the speech and toward what is in their hands, no matter what you may say. Use handouts only in the following cases:

- In rare instances when no other graphic is possible, as in a room with no equipment of any kind
- In cases when your listeners have requested handouts during the speech
- At the end of the speech, so that material in handouts can help listeners later recollect ideas from your speech

Graphics Use Guideline 3: Maintain Eye Contact

Don't stare at your graphics while you speak. Maintain control of listeners' responses by looking back and forth smoothly from the graphic to faces in the audience. Also, in pointing to the graphic, use the hand closest to it. If instead you use the hand farthest away, you end up crossing your arm over your torso and, thus, turn your neck and head away from the audience.

Graphics Use Guideline 4: Use Your Own Equipment

Murphy's Law always seems to apply when you use another person's audio-visual equipment: Whatever can go wrong, will. For example, a new bulb burns out, there is no extra bulb in the equipment drawer, an extension cord is too short, the screen does not stay down, the client's computer doesn't read your disk—many speakers have experienced these problems. This writer has seen all of these and more. Even if the equipment works, it often operates differently from what you are used to. The only sure way to put the odds in your favor is to carry your own equipment and set it up in advance. However, most of us have to rely on someone else's equipment at least some of the time. Here are some suggestions:

- Find out exactly who will be responsible for providing equipment and contact that person in advance about details.
- Bring some easy-to-carry backup supplies with you—an extension cord, and overhead projector bulb, felt-tip markers, and chalk, for example.
- Bring handout versions of your graphics to use as a backup.

In short, avoid putting yourself in the position of having to apologize. Plan well.

SPECIFIC GUIDELINES FOR EIGHT GRAPHICS

Speech graphics come in many forms. This section includes guidelines for producing specific graphics that appear in all types of formats—from low-tech visual aids such as posters to high-tech graphics such as PowerPoint slides. The design principles and guidelines presented here transcend issues of technology. The following eight types of graphics are covered:

1. Tables
2. Pie charts
3. Bar charts
4. Line charts
5. Schedule charts
6. Flowcharts
7. Organization charts
8. Technical drawings

TABLES

Tables present the audience with raw data, usually in the form of numbers but sometimes in the form of words. Tables are classified as either informal or formal:

- **Informal tables:** Limited data arranged in the form of either rows or columns
- **Formal tables:** Complex data arranged in a grid, always with both horizontal rows and vertical columns

The following five guidelines help you design and position tables within the speech text.

Table Guideline 1: Use Informal Tables When Possible

Documents provide the reader with ample time to examine detailed graphics. Speeches, however, don't give listeners an opportunity for careful review. Informal tables usually are preferred because they include only rows or columns, not both. As Figure 5–1 shows, an informal table usually has (1) no table number or title, and (2) few, if any, headings for rows or columns.

Table Guideline 2: Use Formal Tables for Complex Data

A formal table can be used if all data on it are important for your presentation. Do *not* use a formal table if most of the data will be ignored—it would be more of a distraction than an aid. When you use formal tables, do the following:

- Extract important data from the table and highlight them in the speech.
- Leave the table up long enough for important information to be absorbed; then if necessary bring it back if it is needed again.
- Use color to separate important information from surrounding detail.

Table Guideline 3: Use Plenty of White Space

Used around and within tables, white space guides the eye through a table much better than do black borders. Avoid putting densely drawn black boxes

Our project in Alberta, Canada, will involve engineers, technicians, and salespeople from three other offices, in these numbers:

San Francisco Office		45
St. Louis Office		34
London Office		6
	Total	85

FIGURE 5–1
Example of informal table.

around tables. Instead, leave 1 inch more of white space than you would normally leave around text and let it act as a frame.

Table Guideline 4: Follow the Design Conventions for Complex Tables

Figure 5–2 shows a typical formal table. It satisfies the overriding goal of being clear and self-contained. To achieve that objective in your tables, follow these guidelines:

1. **Titles and numberings:** Give a title to each formal table, and place title and number above the table. Number each table if the speech contains two or more tables.
2. **Headings:** Create short, clear headings for all columns and rows.
3. **Abbreviations:** Include in the headings any necessary abbreviations or symbols, such as lb or %. Explain abbreviations and define terms if listeners may need such assistance.
4. **Numbers:** Round off numbers when possible, for ease of reading. Also, align multidigit numbers on the right edge, or at the decimal when shown.
5. **Notes:** Place any explanatory headnotes either between the title and the table (if the notes are short) or at the bottom of the table.
6. **Sources:** Place any source references below the footnote and make reference to the source during your speech.
7. **Caps:** Use uppercase and lowercase letters, rather than full caps.

Remember, however, that all writing should be readable by the entire audience. If it is not, remove it from the table during preparation and give the information (such as sources and notes) in your speech.

Table Guideline 5: Pay Special Attention to Cost Data

Most readers prefer to have financial information placed in tabular form. Given the importance of such data, edit cost tables with great care. Devote extra attention to these two issues:

- Placement of decimals in costs
- Correct totals of figures

TABLE 22: Employee Retirement Fund

Investment Type	Book Value	Market Value	% of Total Market Value
Temporary Securities	$ 434,084	434,084	5.9%
Bonds	3,679,081	3,842,056	52.4
Common Stocks	2,508,146	3,039,350	41.4
Mortgages	18,063	18,063	.3
Real Estate	1,939	1,939	nil
Totals	$6,641,313	$7,335,492	100.0%

FIGURE 5–2
Example of formal table.

PIE CHART

Pie charts show approximate relationships between the parts and the whole. Their simple circles with clear labels can provide simplicity within even the most complicated speech. Yet the design simplicity of the circle means that pie charts are *not* the best choice when you need to show detailed information or changes over time. Following are basic guidelines for constructing pie charts.

Pie Chart Guideline 1: Use No More than Six or Seven Divisions

To make pie charts work well, limit the number of pie pieces to no more than six or seven. In fact, the fewer segments the better. This approach lets the audience grasp major relationships without having to wade through the clutter of tiny divisions that are difficult to see. In Figure 5–3, for example, the company's client can readily see that most of the project staff will come from three offices close to the project site in Georgia.

Pie Chart Guideline 2: Move Clockwise from 12:00, from Largest to Smallest Wedge

Audiences prefer pie charts oriented like a clock—with the first wedge starting at 12:00. Move from the largest to the smallest wedge to provide a convenient organizing principle. Make exceptions to this design only for good reason. In Figure 5–3, for example, the last wedge represents a greater percentage than the previous wedge. In this way, it does not break up the sequence the writer wants to establish by grouping the three offices with the three largest percentages of project workers. Another exception over which you have no control is that some software packages will not permit you to begin the pie at 12:00.

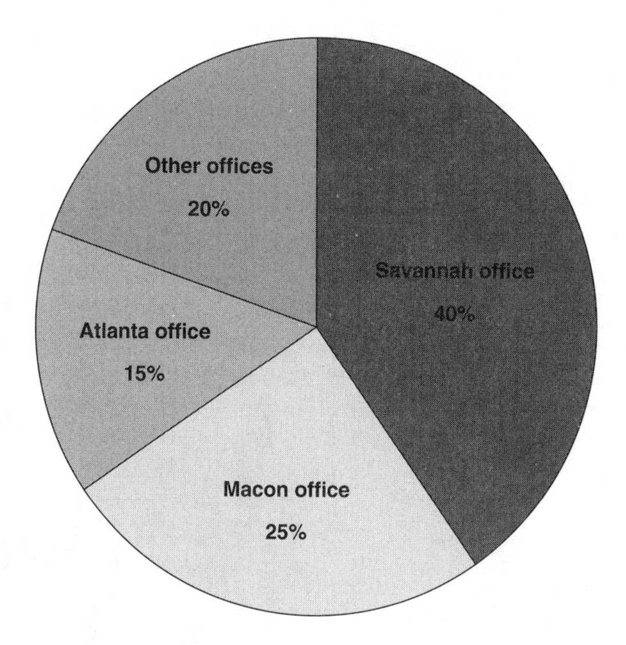

FIGURE 5–3
Example of pie chart.

Pie Chart Guideline 3: Use Pie Charts for Percentages and Money

Pie charts catch the observer's eye best when they represent items divisible by 100—as with percentages and dollars. Figure 5–3 shows percentages; Figure 5–4 shows money. Using the pie chart for money breakdowns is made even more appropriate by the coinlike shape of the chart. *In every case, make sure your percentages or cents add up to 100.*

Pie Chart Guideline 4: Be Creative, but Stay Simple

Figure 5–5 shows the following options for pie charts included in oral presentations:

- Shading a wedge
- Exploding the pie and creating a depth dimension
- Creating a pie within a pie

Be aware that some desktop publishing programs automatically format pie charts using complex backgrounds and shading. However, they may be difficult to read and distort the pieces of the pie. Remember to keep pie charts simple and clean.

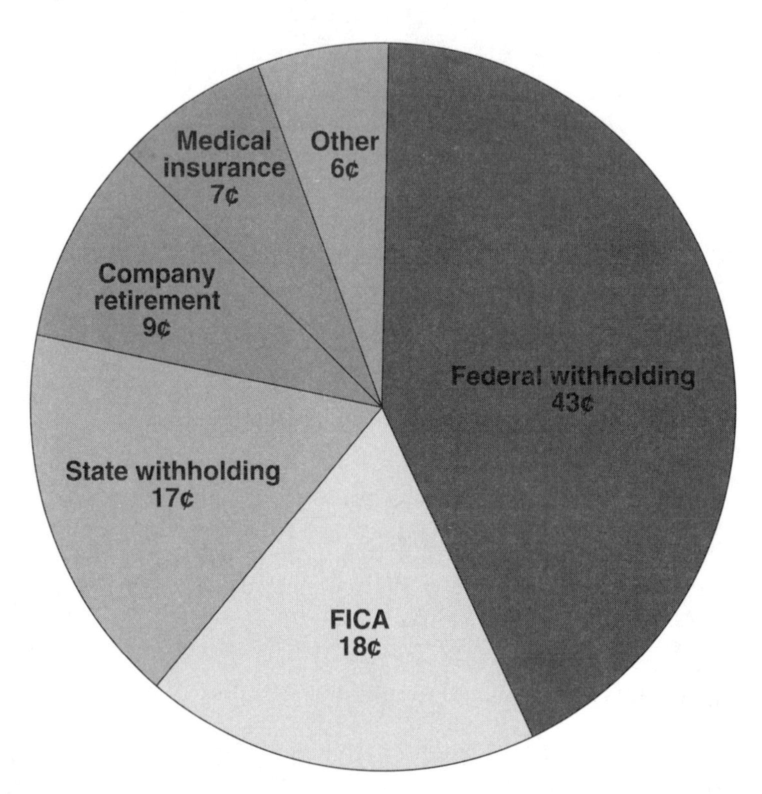

FIGURE 5–4
Example of pie chart.

Pie Chart Guideline 5: Draw and Label Carefully

The most common pie chart errors are (1) wedge sizes that do not correspond correctly to percentages or money amounts and (2) pie sizes that are too small to accommodate the information placed on them. Here are some suggestions for avoiding these mistakes:

- **Pie size:** Make sure the chart occupies enough of the page. On a standard 8 1/2 × 11-inch sheet with only one pie chart, your circle should be about 5 or 6 inches in diameter—large enough to show up clearly when transmitted to a screen during your speech.
- **Labels:** If you use them, place labels either inside the pie or outside, depending on the number of wedges, the number of wedge labels, or the length of the labels. Choose the option that produces the cleanest-looking chart. As always, remember that all wording on a chart must be clearly readable by everyone in the audience. If you can accomplish this purpose, leave labels off the chart and substitute your spoken explanations instead.

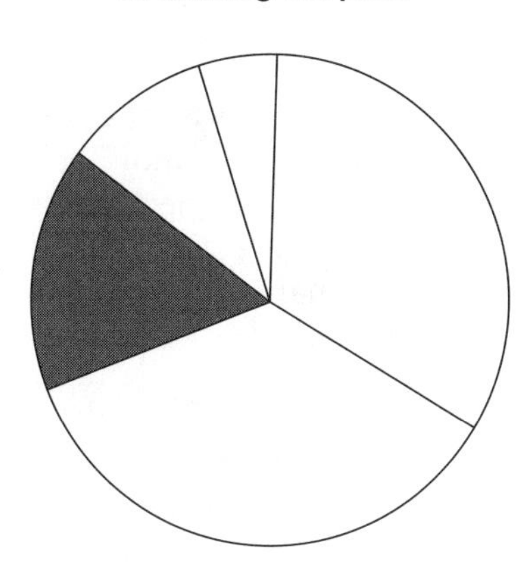

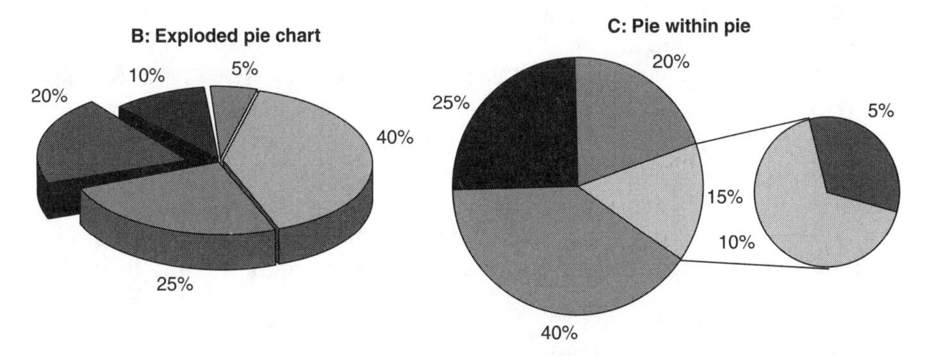

FIGURE 5–5
Pie chart options.

- **Conversion of percentages:** If you draw the pie chart by hand, instead of using a computer program, use a protractor or similar device. One percent of the pie equals 3.6 degrees (3.6 × 100 = 360 degrees in a circle). Thus, you can convert percentages or cents to degrees.

Remember that a pie chart does not reveal fine distinctions very well; it is best used for showing larger differences.

Bar Charts

Like pie charts, bar charts are easily recognized and are often found in newspapers and magazines. However, bar charts can accommodate more technical detail than pie charts without appearing cluttered. Comparisons are provided by means of two or more bars running either horizontally or vertically. Follow these five guidelines to create effective bar charts.

Bar Chart Guideline 1: Limit the Number of Bars

Although bar charts can show more information than pie charts, both types of illustrations have their limits. Bar charts begin to break down when there are so many bars that information is not easily grasped. The maximum number of bars can vary according to chart size, of course. However, remember that a bar chart's impact on the audience is enhanced by reducing the number of bars.

Bar Chart Guideline 2: Show Comparisons Clearly

Bar lengths should be varied enough to show comparisons quickly and clearly. Avoid using bars that are too close in length, for then the audience must study the chart for a long time before grasping its meaning. The chart should convey an immediate visual impact.

Avoid the opposite tendency of using bar charts to show data that are much different in magnitude. To relate such differences, some speakers resort to the dubious technique of inserting "break lines" (two parallel lines) on an axis to reflect breaks in scale (see Figure 5–6). Although this approach at least reminds your audience of the breaks, it is deceptive. For example, note that Figure 5–6 provides no visual demonstration of the relationship between 50 and 2,800. The audience must think about these differences before making sense out of the chart. Here the use of hash marks runs counter to a main goal of graphics—creating an immediate and accurate visual impact.

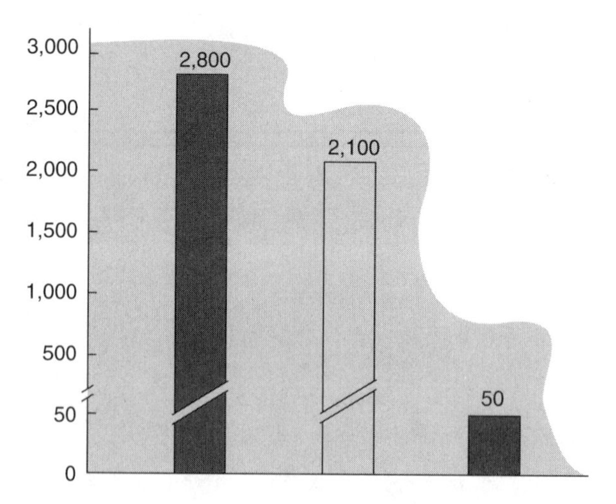

FIGURE 5–6
Misleading bar chart with hash marks.

Bar Chart Guideline 3: Keep Bar Widths Equal and Adjust Space Between Bars Carefully

While bar length varies, bar width must remain constant. As for distance between the bars, following are three options (along with examples shown in Figure 5–7):

- **Option A:** Use no space when there are close comparisons or many bars, so that differences are easier to grasp.
- **Option B:** Use equal space, but less than the bar width, when bar height differences are great enough to be seen in spite of the distance between the bars.
- **Option C:** Use variable space in gaps between some bars to reflect gaps in the data.

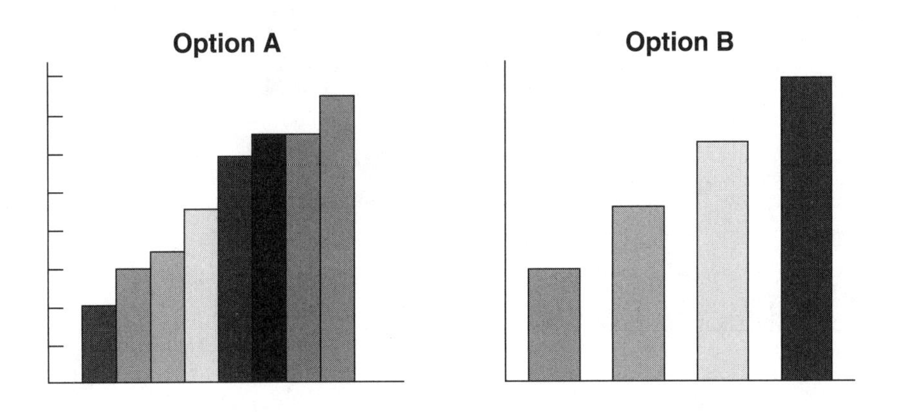

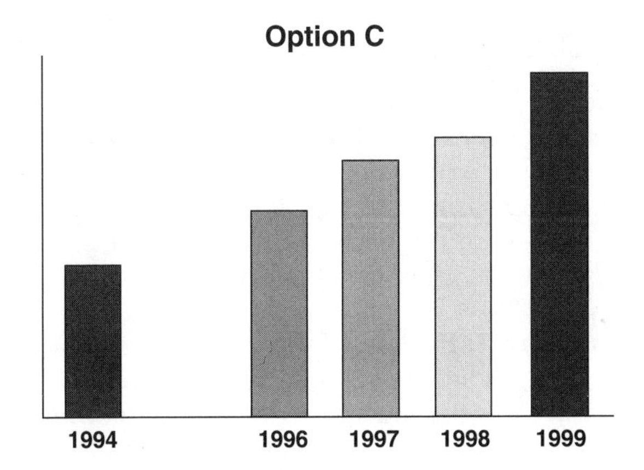

FIGURE 5–7
Bar chart variations.

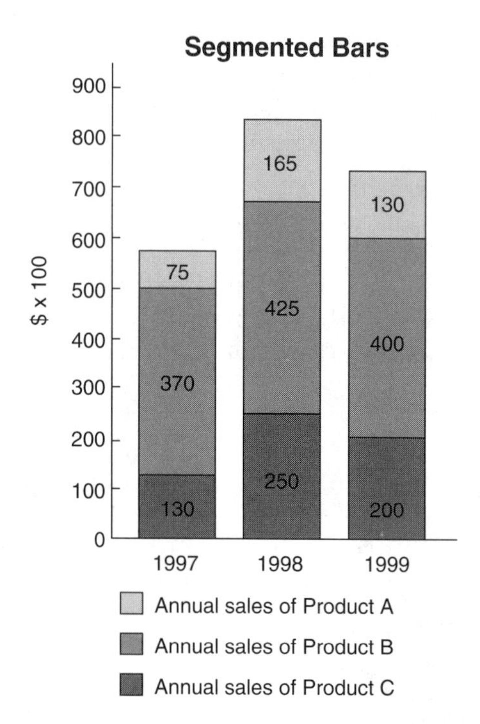

FIGURE 5–8
Bar chart variations for multiple trends.

Bar Chart Guideline 4: Arrange the Order of the Bars Carefully

The arrangement of bars reveals much of the chart's meaning to the audience. Here are two common approaches:

- **Sequential:** Used when the progress of the bars shows trends—for example, the increasing number of environmental projects in the last five years
- **Ascending or descending order:** Used when you want to make a point by the rising or falling of the bars—for example, the total sales of your firm's six offices for 2000, from lowest total to highest total

Bar Chart Guideline 5: Be Creative

Figures 5–8 and 5–9 show two bar chart variations that help display multiple trends. The segmented bars in Figure 5–8 produce four types of information: the total sales (A + B + C) and the individual sales for A, B, and C. The grouped bars in Figure 5–9 show the individual sales trends for D, F, and G, along with a comparison of all three by year. Note that the amounts are written above the bars to highlight comparisons.

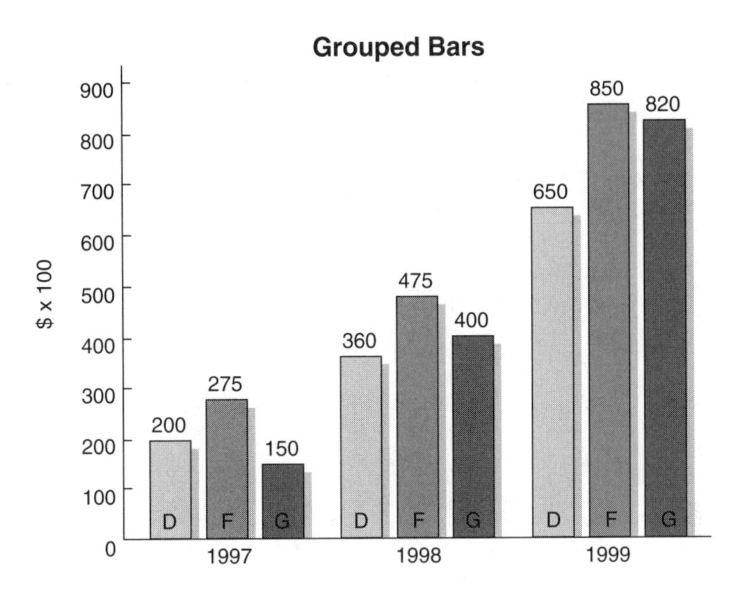

FIGURE 5–9
Bar chart vairations for multiple trends.

LINE CHARTS

Line charts are a common graphic in speeches and documents. Almost every newspaper contains a few charts covering topics such as stock trends, car prices, or weather. More than other graphics, line charts telegraph complex trends immediately.

They work by using vertical and horizontal axes to reflect quantities of two different variables. The vertical (or *y*) axis usually plots the dependent variable; the horizontal (or *x*) axis usually plots the independent variable. (The dependent variable is affected by changes in the independent variable.) Lines then connect points that have been plotted on the chart. When drawing line charts, follow these five main guidelines.

Line Chart Guideline 1: Use Line Charts for Trends

The audience is influenced by the direction and angle of the chart's line, so take advantage of this persuasive potential. In Figure 5–10, for example, the speaker wants to show the feasibility of adopting a new medical plan for an organization. Including a line chart gives immediate emphasis to the most important issue—the effect the new plan will have on stabilizing the firm's medical costs.

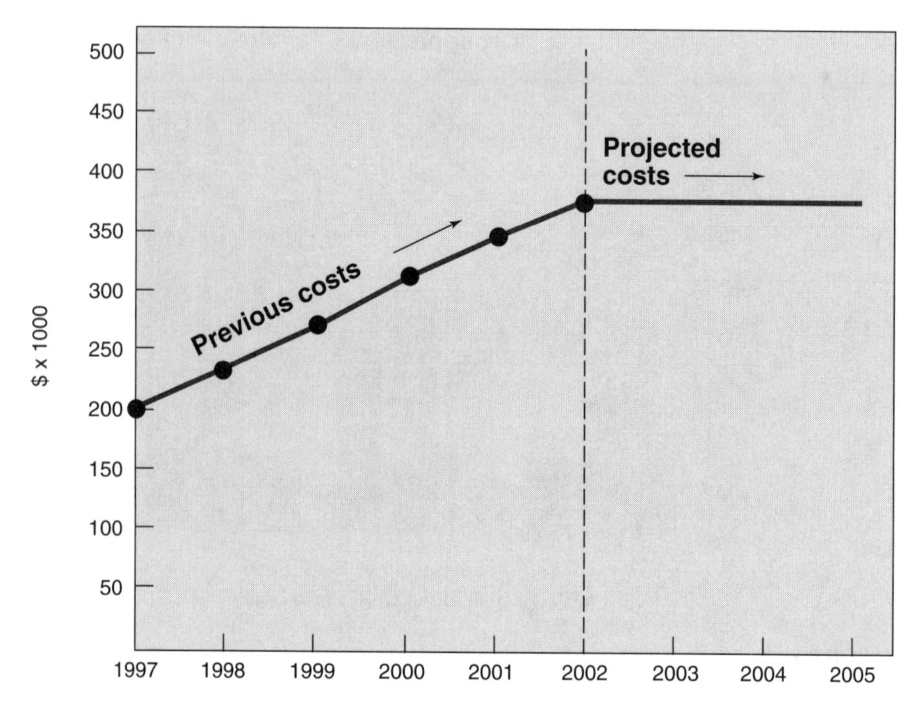

FIGURE 5–10
Line chart.

Line Chart Guideline 2: Locate Line Charts with Care

Given their strong impact, line charts can be especially useful as attention grabbers. Consider using them (1) at the beginning of the speech to engage interest, (2) at the beginning of sections that describe trends, and (3) in your conclusions to reinforce a major point of the speech.

Line Chart Guideline 3: Strive for Accuracy and Clarity

Like bar charts, line charts can be misused or poorly constructed. Be sure the line or lines on the graph accurately reflect the data from which you have drawn. Also, select a scale that does not mislead readers with visual gimmicks. Here are some specific suggestions to keep line charts accurate and clear:

- Start all scales from zero to eliminate the possible confusion of breaks in amounts (see Bar Chart Guideline 2).
- Select a vertical-to-horizontal ratio for axis lengths that is pleasing to the eye (three vertical to four horizontal is common).
- Make chart lines as thick as, or thicker than, the axis lines.
- Use shading under the line when it will make the chart more readable.

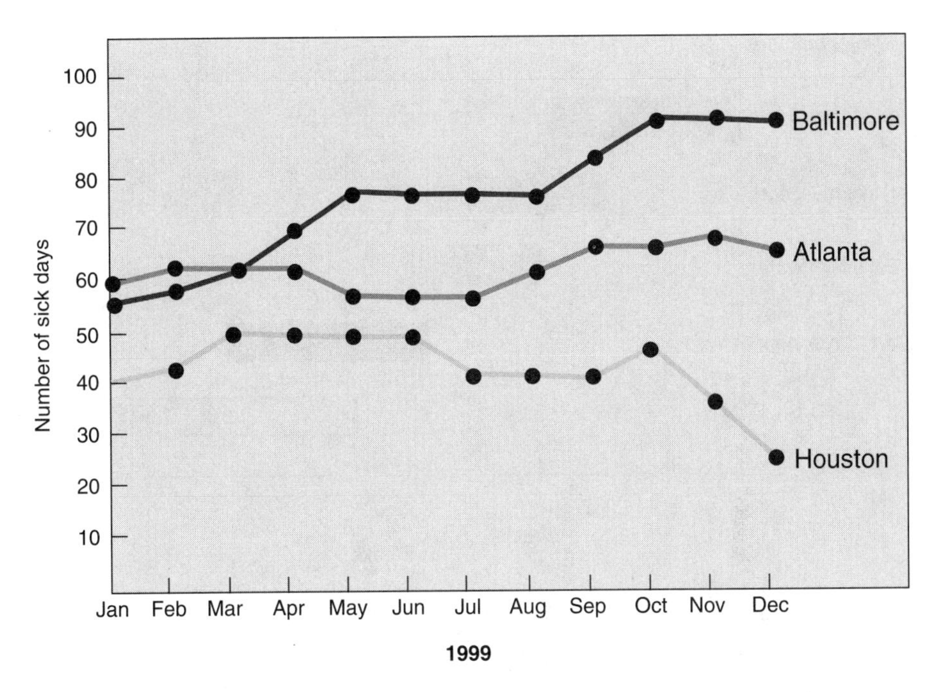

FIGURE 5–11
Line chart.

Line Chart Guideline 4: Do Not Place Numbers on the Chart

Line charts derive their main effect from the simplicity of lines that show trends. Avoid cluttering the chart with a lot of numbers that only detract from the visual impact.

Line Chart Guideline 5: Use Multiple Lines with Care

Like bar charts, line charts can show multiple trends. Simply add another line or two. If you place too many lines on one chart, however, you risk confusing the audience with too much data. Use no more than four or five lines on a single chart (see Figure 5–11).

Schedule Charts

Many speeches, especially those reporting on proposals and feasibility studies, include a special kind of chart that shows readers when activities will be accomplished. The schedule chart highlights tasks and times mentioned in the speech text and usually includes these parts (see Figure 5–12):

- Vertical axis, which lists the various parts of the project, in sequential order
- Horizontal axis, which registers the appropriate time units
- Horizontal bar lines or separate markers, which show the starting and ending times for each task

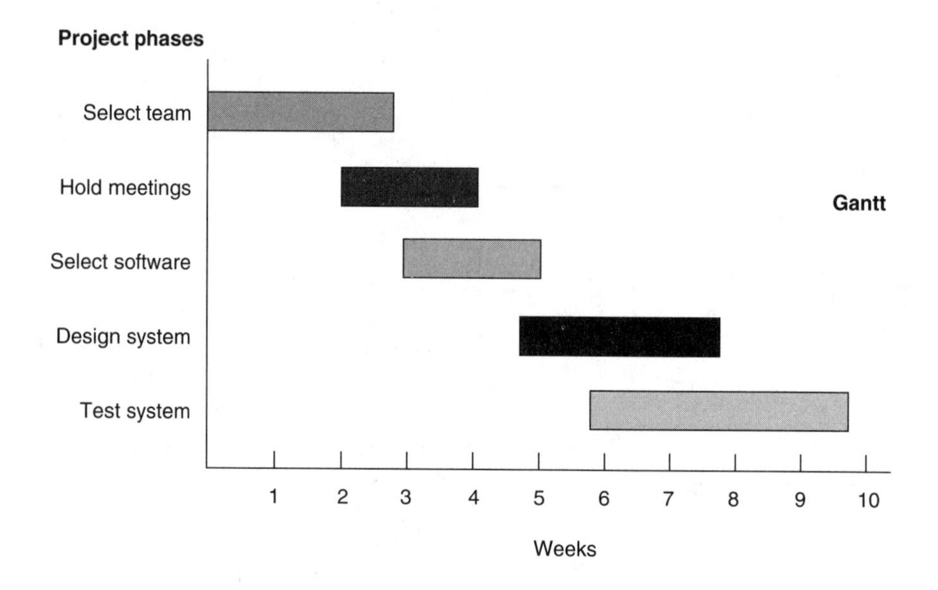

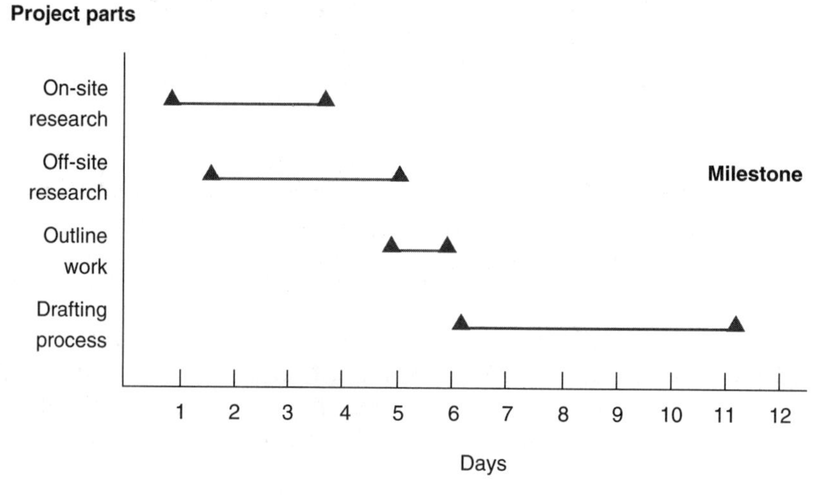

FIGURE 5–12
Schedule charts.

Use the following guidelines for constructing schedule charts included in speeches related to proposals, feasibility studies, and other documents.

Schedule Chart Guideline 1: Include Only Main Activities

Keep the audience focused on no more than ten or fifteen main activities. If more detail is needed, construct a series of schedule charts linked to a main "overview" chart.

Schedule Chart Guideline 2: List Tasks in Sequence from the Top

As shown in Figure 5–12, the convention is to list activities from the top to the bottom of the vertical axis. Thus, the observer's eyes move from the top left to the bottom right of the page, the most natural flow for many people.

Schedule Chart Guideline 3: Run All Labels in a Horizontal Direction

If audience members cannot easily read labels, they may lose interest.

Schedule Chart Guideline 4: Create New Formats when Needed

Figure 5–12 shows only two common types of schedule charts; you can devise your own hybrid form when it suits your purposes. Your goal is to find the simplest format for showing your audience when a product will be delivered or a service completed, for example.

Schedule Chart Guideline 5: Be Realistic about the Schedule

Schedule charts can come back to haunt you if you don't include feasible deadlines. As you set dates for activities, be realistic about the likely time something can be accomplished. Your managers and clients understand delays caused by weather, equipment breakdowns, and other unforeseen events. However, they will be less charitable about schedule errors that result from sloppy planning.

FLOWCHARTS

Flowcharts tell a story about a process, usually stringing together a series of boxes and other shapes that represent separate activities (see Figure 5–13). Because they have a reputation for being hard to read, you need to take extra care in designing them. These five guidelines will help.

Flowchart Guideline 1: Present Only Overviews

Your audience usually wants a flowchart that gives only a capsule version of the process, not all the details. Reserve your list of particulars for the text of the speech or a handout provided *after* the speech concludes.

Flowchart Guideline 2: Limit the Number of Shapes

Flowcharts rely on rectangles and other shapes to relate a process—in effect, to tell a story. Different shapes represent different types of activities. This variety helps in describing a complex process, but it can also produce confusion. For the sake of clarity and simplicity, limit the number of different shapes in your flowcharts.

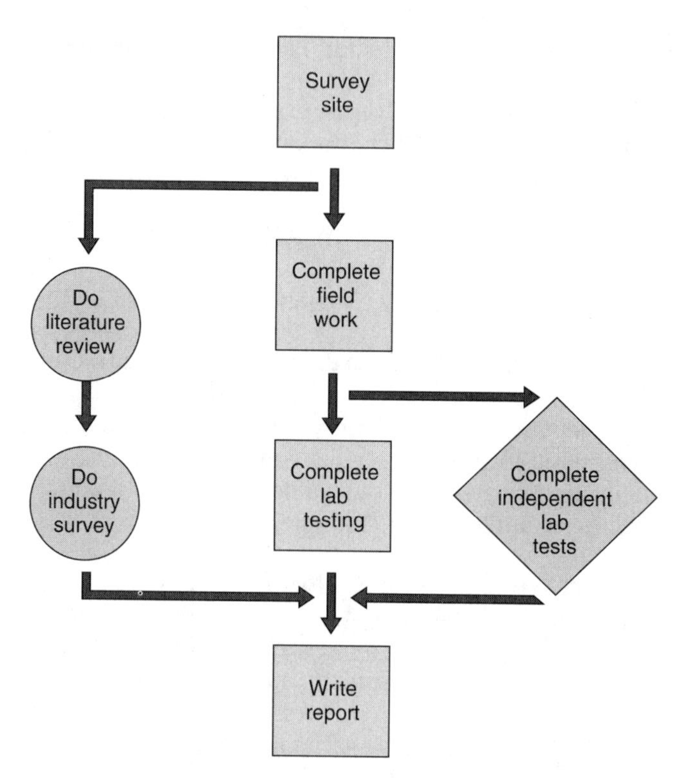

FIGURE 5–13
Flowchart.

Flowchart Guideline 3: Provide a Legend When Necessary

Simple flowcharts often need no legend. The few shapes on the chart may already be labeled by their specific steps. When charts get more complex, however, you can include a legend that identifies the meaning of each shape.

Flowchart Guideline 4: Run the Sequence from Top to Bottom or from Left to Right

Long flowcharts may cover an entire page. Yet they should always show some degree of uniformity by assuming either a basically vertical or horizontal direction.

Flowchart Guideline 5: Label All Shapes Clearly

In addition to a legend that defines meanings of different shapes, the chart usually includes a label for each individual shape or step. Follow one of these approaches:

- Place the label inside the shape.
- Place the label immediately outside the shape.
- Put a number on each shape and place a legend for all numbers in another location.

ORGANIZATION CHARTS

Organization charts reveal the structure of an organization—people, positions, or work units. The challenge in producing this graphic is to make certain the arrangement of information reflects the organization accurately.

Organization Chart Guideline 1: Use Linear "Boxes" to Emphasize Hierarchy

This traditional format uses rectangles connected by lines to represent some or all of the positions in an organization (see Figure 5–14). Because high-level positions usually appear at the top of the chart, where the attention of most readers is focused, this design tends to emphasize upper management.

Organization Chart Guideline 2: Connect Boxes with Lines

Solid lines show direct reporting relationships; dotted lines show indirect or staff relationships (see Figure 5–14).

Organization Chart Guideline 3: Use a Circular Design to Avoid Emphasis on Hierarchy

This arrangement of concentric circles gives more visibility to workers outside upper management. These are often the technical workers most deeply involved in the details of a project. For example, Figure 5–15 draws attention to the project engineers perched on the chart's outer ring.

Organization Chart Guideline 4: Use Varied Shapes Carefully

Like flowcharts, organization charts can use different shapes to indicate different levels or types of jobs. However, beware of introducing more complexity than you need. Use more than one shape only if you are convinced this approach is needed to convey meaning to the audience.

Organization Chart Guideline 5: Be Creative

When standard forms will not work, create new ones. For example, Figure 5–16 uses an organization chart as the vehicle for showing the lines of responsibility in a specific project.

TECHNICAL DRAWINGS

Technical drawings are important tools for companies that produce or use technical products. These drawings can accompany speeches that cover

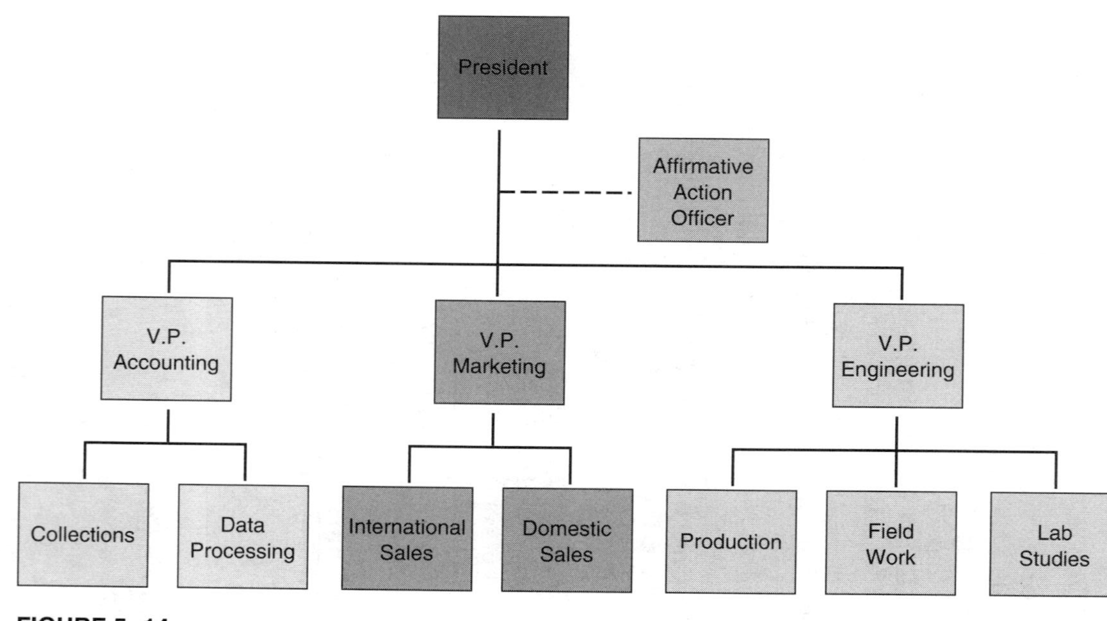

FIGURE 5–14
Basic organization chart.

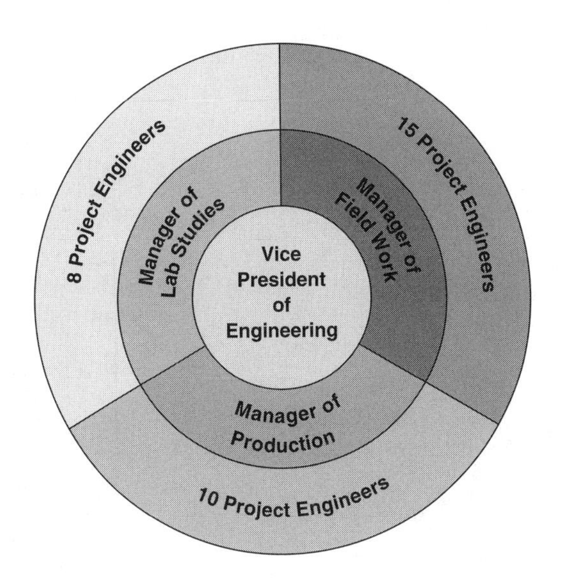

FIGURE 5–15
Concentric organization chart.

instructions, reports, and proposals, for example. They are preferred over photographs when specific views are more important than photographic detail. Whereas all drawings used to be produced mainly by hand, now they are usually created by CAD (computed-assisted design) systems. Follow these guidelines for producing technical drawings that complement your text.

Drawing Guideline 1: Choose the Right Amount of Detail

Keep drawings in oral presentations as simple as possible. Use only the level of detail that serves the purpose of your speech and satisfies your listener's needs. For example, Figure 5–17 will be used in a speech about maintaining home heating systems. Its intention is to focus on one part of the thermostat—the lever and attached roller. Completed on a CAD system, this drawing presents an exploded view so that the location of the arm can be seen easily.

Drawing Guideline 2: Label Parts Appropriately for a Speech

A common complaint about drawings used in speeches is that they include too much detail. Use only labels that are relevant to your speech. Most important, make sure they can be read by every member of the audience and do not detract from the importance of the drawing itself. The simple labeling in Figure 5–17 fulfills these objectives.

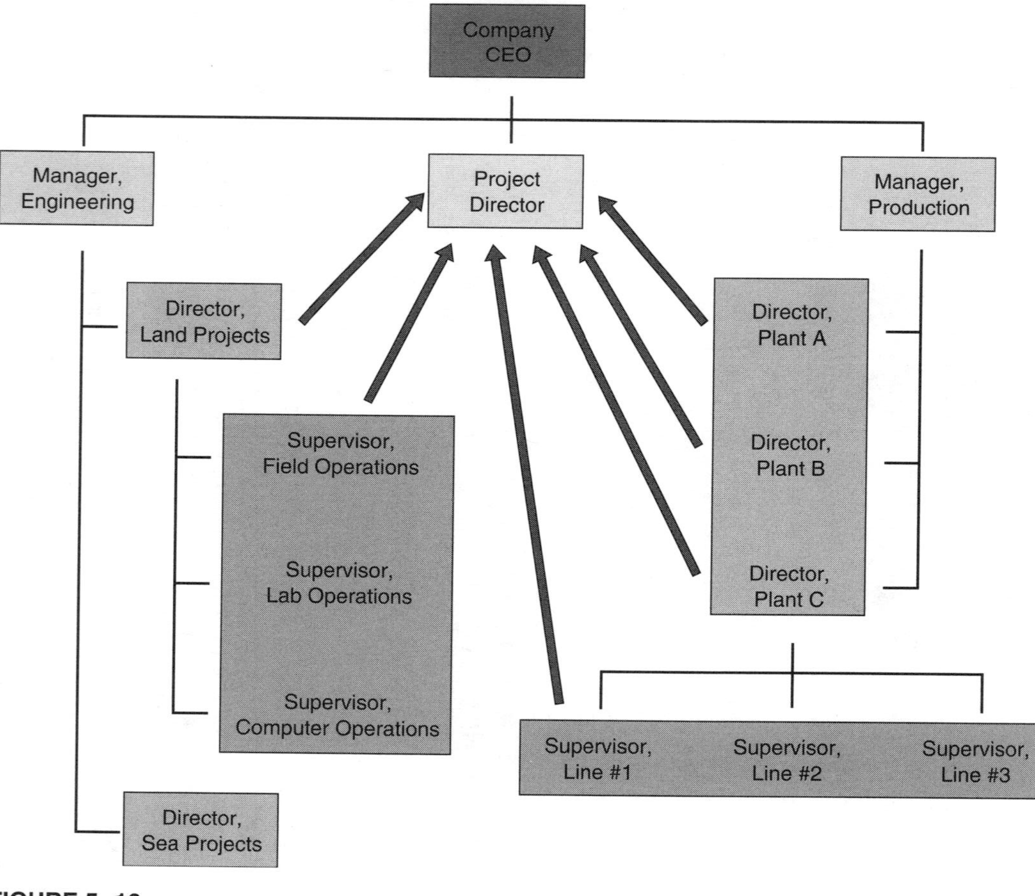

FIGURE 5–16
Organization chart focusing on project.

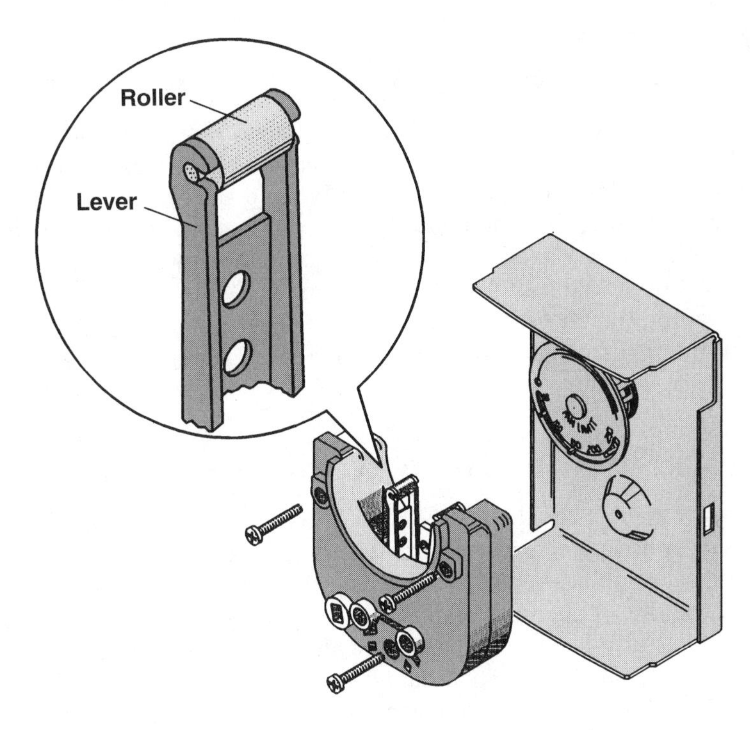

FIGURE 5–17
Technical drawing.

Drawing Guideline 3: Choose the Most Appropriate View

As already noted, illustrations—unlike photographs—permit you to choose the level of detail needed. In addition, drawings offer you a number of options for perspective or view:

- **Exterior view:** Shows surface features with either a two- or three-dimensional appearance.
- **Cross-sectional view:** Shows a "slice" of the object so that interiors can be viewed.
- **Exploded view:** Shows relationship of parts to each other by "exploding" the mechanism—see Figure 5–17.
- **Cutaway view:** Shows inner workings of an object by removing part of the exterior.

Drawing Guideline 4: Use Legends When There Are Many Parts

In complex drawings, avoid cluttering the illustration with many labels. You may want to place all labels in one easy-to-find legend. Other times the need for readability will require that you eliminate labels on the drawing used in the speech, perhaps providing them on a handout given out afterwards.

MISUSE OF GRAPHICS

Technology has revolutionized the world of graphics by placing sophisticated tools in the hands of many users. Yet this largely positive event has its dark side. You may have seen some graphics that—in spite of their professional appearance—distort data and misinform the audience. The previous sections of this chapter have established principles and guidelines to help you avoid such distortion and misinformation. This last section shows what can happen to graphics when sound design principles are not applied. It also offers suggestions for correctly employing that increasingly popular form of speech graphics—PowerPoint.

DESCRIPTION OF THE GRAPHICS PROBLEM

The popular media provide one window into the problem of faulty graphics. One observer used newspaper reports about the October 19, 1987 stock market plunge as an indication of the problem. Writing in *Aldus Magazine,* Daryl Moen noted that 60 percent of U.S. newspapers included charts and other graphics about the market drop the day after it occurred. Moen's study revealed that one out of eight had data errors, and one out of three distorted the facts with visual effects. That startling statistic suggests that faulty illustrations are a genuine problem (Daryl Moen, "Misinformation Graphics," *Aldus Magazine,* January/February 1990, p. 64).

Edward R. Tufte analyzes graphics errors in more detail in his excellent work, *The Visual Display of Quantitative Information.* In setting forth his main principles, Tufte notes that "graphical excellence is the well designed presentation of interesting data—a matter of substance of statistics, and of design." He further contends that graphics must "give to the viewer the greatest number of ideas in the shortest time with the least ink in the smallest space" (Edward R. Tufte, *The Visual Display of Quantitative Information,* Cheshire, CT: Graphics Press, 1983, p. 51).

One of Tufte's main criticisms is that charts are often disproportional to actual differences in the data presented. The next section shows specific ways that this error has worked its way into contemporary graphics.

EXAMPLES OF DISTORTED GRAPHICS

There are probably as many ways to distort graphics as there are graphic types. This section gets at the problem of misrepresentation by showing several examples and describing the errors involved. Each example fails to represent data accurately.

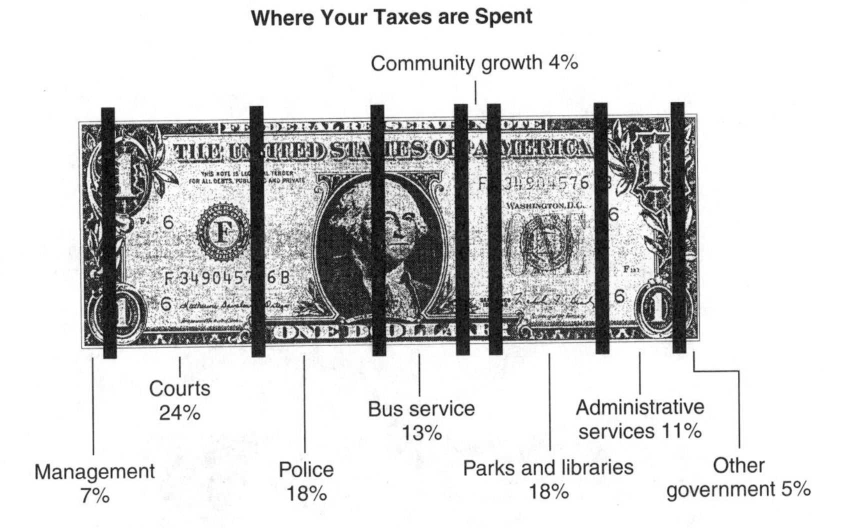

FIGURE 5–18
Chartjunk that confuses the audience.

Example 1: "Chartjunk" that Confuses the Reader

Figure 5–18 concludes a report from a government agency to its citizens. Whereas the dollar backdrop is meant to reinforce the topic—that is, the use to which tax funds are put—in fact, it impedes communication. The audience cannot quickly see comparisons.

At the very least, the expenditures should have been placed in sequence, from least to greatest percentage or vice versa. Even with this order, however, one could argue that the dollar bill is a piece of what Edward Tufte calls "chartjunk," which fails to display the data effectively.

Example 2: Confusing Pie Charts

The pie chart shown in Figure 5–19 (1) fails to move in a largest-to-smallest, clockwise sequence, (2) includes too many divisions, many of which are about the same size and are difficult to distinguish, and (3) includes wording that would be difficult for the audience to see. Figure 5–20 negates the value of the pie chart by assuming an oblong shape, rather than a circle. This distortion can make it difficult for the reader to distinguish among sections that are similar in size, such as Sections 1 and 2. Preferably, the pie chart should be a perfect circle and should move in large-to-small sequence from the 12:00 position.

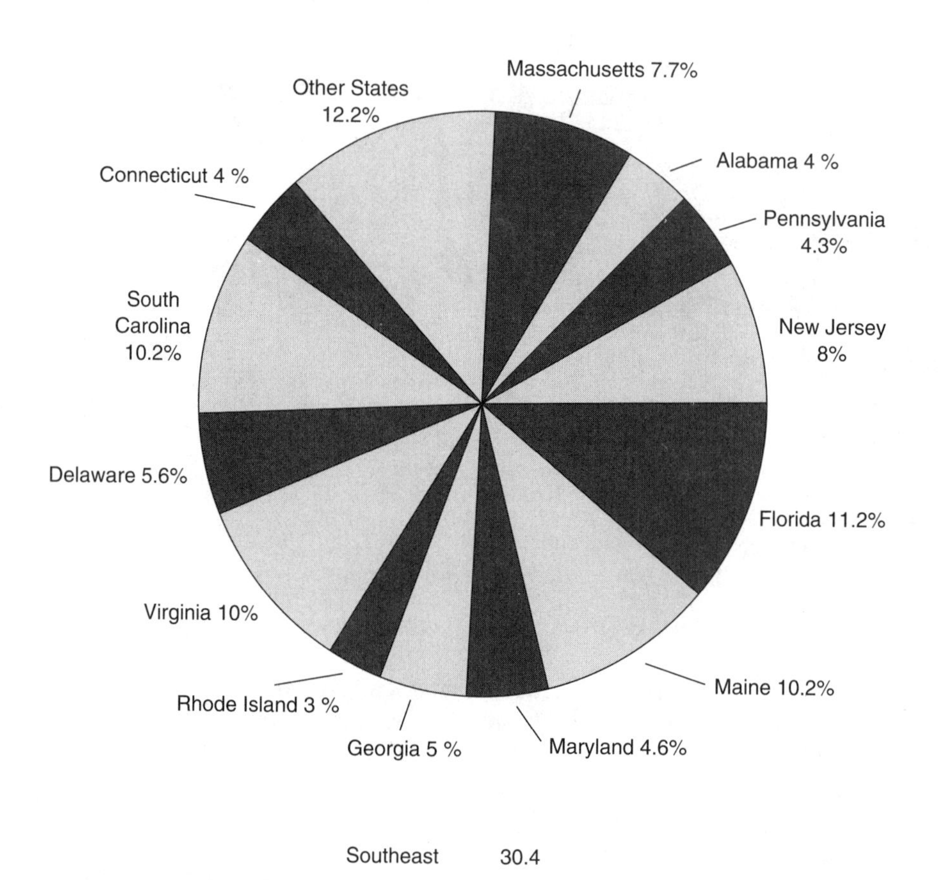

FIGURE 5–19
Confusing pie chart.

Special Issues Related to PowerPoint

This chapter has focused on selecting appropriate graphics and designing them properly. Our province has not included specific technologies used to produce graphics. However, one particular technology—PowerPoint—is so pervasive in speeches of all kinds that it deserves special mention.

Microsoft's PowerPoint permits you to use your computer to create a wide variety of graphics, which then can be incorporated into a speech or written report. For a speech, generally you would need a laptop connected to a videoprojector in the room where you are speaking. Then while you are speaking, you would move through the electronic slide show stored on

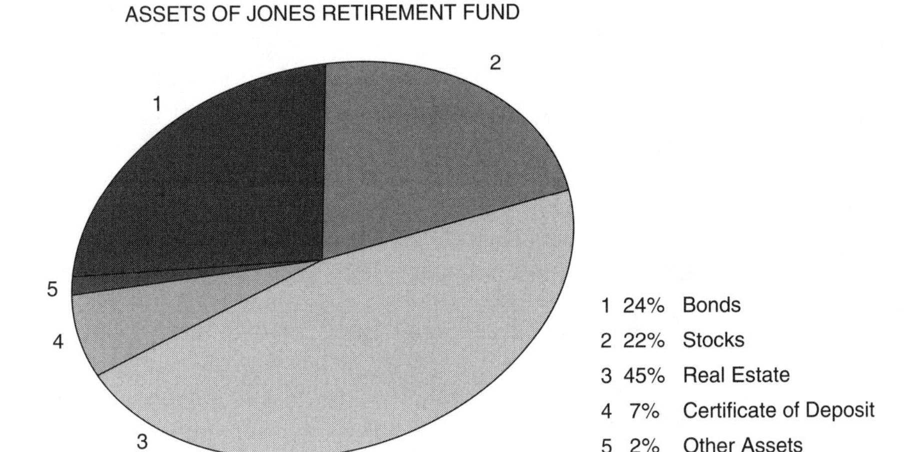

FIGURE 5–20
Confusing pie chart.

your computer. This section includes a few basic guidelines for using this tool properly.

PowerPoint Guideline 1: Spend Time with the Tutorial

Microsoft provides tutorials that help you decide how, when, and where to use PowerPoint slides. You will soon discover that the software offers a dizzying array of charts, pictures, clip art, voice narration, and more. If you spend some front-end time getting to know the capabilities of the package, you will be more likely to make good choices later.

PowerPoint Guideline 2: Start Simple and Stay That Way

With so many design options available, some PowerPoint users tend to select a wide variety of colors, shapes, screen movements, and other options. That may be a mistake. It is usually best to stay with simple, clear principles of design so that your audience will focus on the material presented, not on the "bells and whistles" of the delivery. The KISS principle mentioned earlier—"Keep It Short and Simple"—applies to PowerPoint presentations just as it does to speech text and other graphics.

PowerPoint Guideline 3: Avoid Excessive Use of Text

One of the most common mistakes with PowerPoint is simply to generate a good deal of text on the screen. Don't use PowerPoint technology as an excuse to present the kind of excessive verbiage you would never consider putting on more traditional graphics, such as overhead transparencies.

PowerPoint Guideline 4: Do Not Read Directly from the Slides

The preparation error noted in the previous guideline often leads to the delivery error noted here. PowerPoint does not relieve you of a basic obligation of all speech graphics—that is, to maintain eye contact and to present material in an engaging fashion. An especially irksome problem with PowerPoint is the tendency of some speakers to simply read text off the screen. If your slides do include wording, make them brief and use them for listing "talking points" you will expand upon as you speak to the audience, not to the screen.

PowerPoint Guideline 5: Check Out Equipment Before the Presentation

More than one presenter has shown up with a PowerPoint-filled laptop, only to find that the room does not include the right connecting cords for the videoprojector. This and other technical headaches can be eliminated if you or a trusted assistant check equipment ahead of time. As mentioned earlier, it's also wise to carry as much of your own equipment as possible.

PowerPoint Guideline 6: Bring a Hard Copy of Slides

If something does go wrong with the hardware or software, you will be glad you brought along a hard copy of the PowerPoint slides. They can be converted quickly to overhead transparencies or handouts.

PowerPoint Guideline 7: Don't Let the Technology Take Over

As always, your audience dictates the degree to which you should rely on graphics. However, remember that most people still value the human connection and prefer to hear a great speech reinforced by effective visuals, rather than a high-tech visual display that eclipses a token speech.

CHAPTER SUMMARY

This chapter shows you how to incorporate various types of graphics into oral presentations. It begins by listing some general guidelines that apply to the preparation and use of all speech graphics, such as the need to respond to graphics preferences of the audience and to plan graphics early. Then it includes suggestions for using and producing eight common types of visuals: tables, pie charts, bar charts, line charts, schedule charts, flowcharts, organization charts, and technical drawings. The chapter concludes with a section on the misuse of graphics in oral presentations. Examples of "chartjunk" and other confusing visuals are followed by guidelines for the appropriate use of PowerPoint.

EXERCISES

Assume that all of the assignments that follow are intended to support related speeches. Thus, you should follow all of the general guidelines for speech graphics at the beginning of the book, as well as specific guidelines for the particular type of graphic you are creating.

1. **Pie, Bar, and Line Charts**
   Figure 5–21 shows total U.S. energy production and consumption from 1960 through 1987, while also breaking down both into the four categories of coal, petroleum, natural gas, and "other." Use those data to complete the following charts:
   - A pie chart that shows the four groupings of energy consumption in 1987
   - A bar chart that shows the trend in total consumption during these four years: 1960, 1965, 1970, and 1975
   - A segmented bar chart that shows the total energy production, and four percent-of-production subtotals, for 1960 and 1980
   - A single-line chart showing energy production from 1965 through 1975
   - A multiple-line chart that contrasts the coal, petroleum, and natural gas percent-of-production for any ten-year span on the table

2. **Schedule Charts**
   Using any options discussed in this chapter, draw a schedule chart that reflects your work on one of the following:
   - A project at works
   - A project at school
   - A project within your community
3. **Flowcharts**
   Select a process with which you are familiar because of work, school, home, or other interests. Then draw a flowchart that outlines the main activities involved in this process.
4. **Organization Charts**
   Select an organization with which you are familiar, or one about which you can find information. Then construct a linear flowchart that would help an outsider understand the management structure of all or part of the organization.
5. **Technical Drawings**
   Drawing freehand or using computer-assisted design, produce a simple technical drawing of an object with which you are familiar.
6. **Tables**
   Using the map in Figure 5–22, construct an informal table correlating the five main groupings with the number of states in each.

No. 926. Energy Production and Consumption, by Major Source: 1960 to 1987

[Btu=British thermal unit. For Btu conversion factors, see text, section 19. See also *Historical Statistics, Colonial Times to 1970*, series M 76–92]

Year	Total production (quad. Btu)	Percent of production: Coal	Petroleum [1]	Natural gas [2]	Other [3]	Total consumption (quad. Btu)	Percent of consumption: Coal	Petroleum [1]	Natural gas [2]	Other [3]	Consumption/production ratio
1960	41.5	26.1	36.0	34.0	3.9	43.8	22.5	45.5	28.3	3.8	1.06
1961	42.0	24.9	36.2	34.9	4.0	44.5	21.6	45.5	29.1	3.8	1.06
1962	43.6	25.0	35.6	35.1	4.2	46.5	21.3	45.2	29.5	4.0	1.07
1963	45.9	25.8	34.8	35.4	4.0	48.3	21.5	44.9	29.8	3.7	1.05
1964	47.7	26.2	33.9	35.8	4.0	50.5	21.7	44.2	30.3	3.8	1.06
1965	49.3	26.5	33.5	35.8	4.3	52.7	22.0	44.1	29.9	4.0	1.07
1966	52.2	25.8	33.7	36.4	4.1	55.7	21.8	43.8	30.5	3.8	1.07
1967	55.0	25.1	33.9	36.5	4.4	57.6	20.7	43.9	31.2	4.2	1.05
1968	56.8	24.0	34.0	37.6	4.4	61.0	20.2	44.2	31.5	4.1	1.07
1969	59.1	23.5	33.1	38.7	4.8	64.2	19.3	44.1	32.2	4.4	1.09
1970	62.1	23.5	32.9	38.9	4.7	66.4	18.5	44.4	32.8	4.3	1.07
1971	61.3	21.5	32.7	40.5	5.3	67.9	17.1	45.0	33.1	4.8	1.11
1972	62.4	22.6	32.1	39.7	5.6	71.3	16.9	46.2	31.9	5.0	1.14
1973	62.1	22.5	31.4	39.9	6.2	74.3	17.5	46.9	30.3	5.3	1.20
1974	60.8	23.1	30.5	38.9	7.4	72.5	17.5	46.1	30.0	6.5	1.19
1975	59.9	25.0	29.6	36.8	8.6	70.5	17.9	46.4	28.3	7.4	1.18
1976	59.9	26.1	28.8	36.4	8.6	74.4	18.3	47.3	27.4	7.1	1.24
1977	60.2	26.2	29.0	36.4	8.5	76.3	18.2	48.7	26.1	7.0	1.27
1978	61.1	24.4	30.2	35.6	9.9	78.1	17.6	48.6	25.6	8.1	1.28
1979	63.8	27.5	28.4	35.0	9.1	[4] 78.9	19.1	47.1	26.2	7.7	1.24
1980	64.8	28.7	28.2	34.2	8.9	76.0	20.3	45.0	26.8	7.8	1.17
1981	64.4	28.5	28.2	34.2	9.1	74.0	21.5	43.2	26.9	8.4	1.15
1982	63.9	29.2	28.7	32.0	10.2	70.8	21.6	42.7	26.1	9.6	1.11
1983	61.2	28.2	30.1	30.6	11.2	70.5	22.6	42.6	24.6	10.2	1.15
1984	[5] 65.8	30.0	28.6	30.7	10.7	74.1	23.0	41.9	25.0	10.0	1.13
1985	64.8	29.8	29.3	29.6	11.3	74.0	23.6	41.8	24.1	10.4	1.14
1986	64.3	30.4	28.6	29.0	12.0	74.3	23.2	43.4	22.5	10.9	1.16
1987	64.6	31.2	27.3	29.5	12.0	76.0	23.7	42.9	22.6	10.8	1.18

[1] Production includes crude oil and lease condensate. Consumption includes domestically produced crude oil, natural gas liquids, and lease condensate, plus imported crude oil and products. [2] Production includes natural gas liquids; consumption excludes natural gas liquids. [3] Comprised of hydropower, nuclear power, geothermal energy and other. [4] Represents peak year for U.S. energy consumption. [5] Represents peak year for U.S. energy production.

Source: U.S. Energy Information Administration, *Annual Energy Review*, and unpublished data.

FIGURE 5–21
Reference for exercise 1.

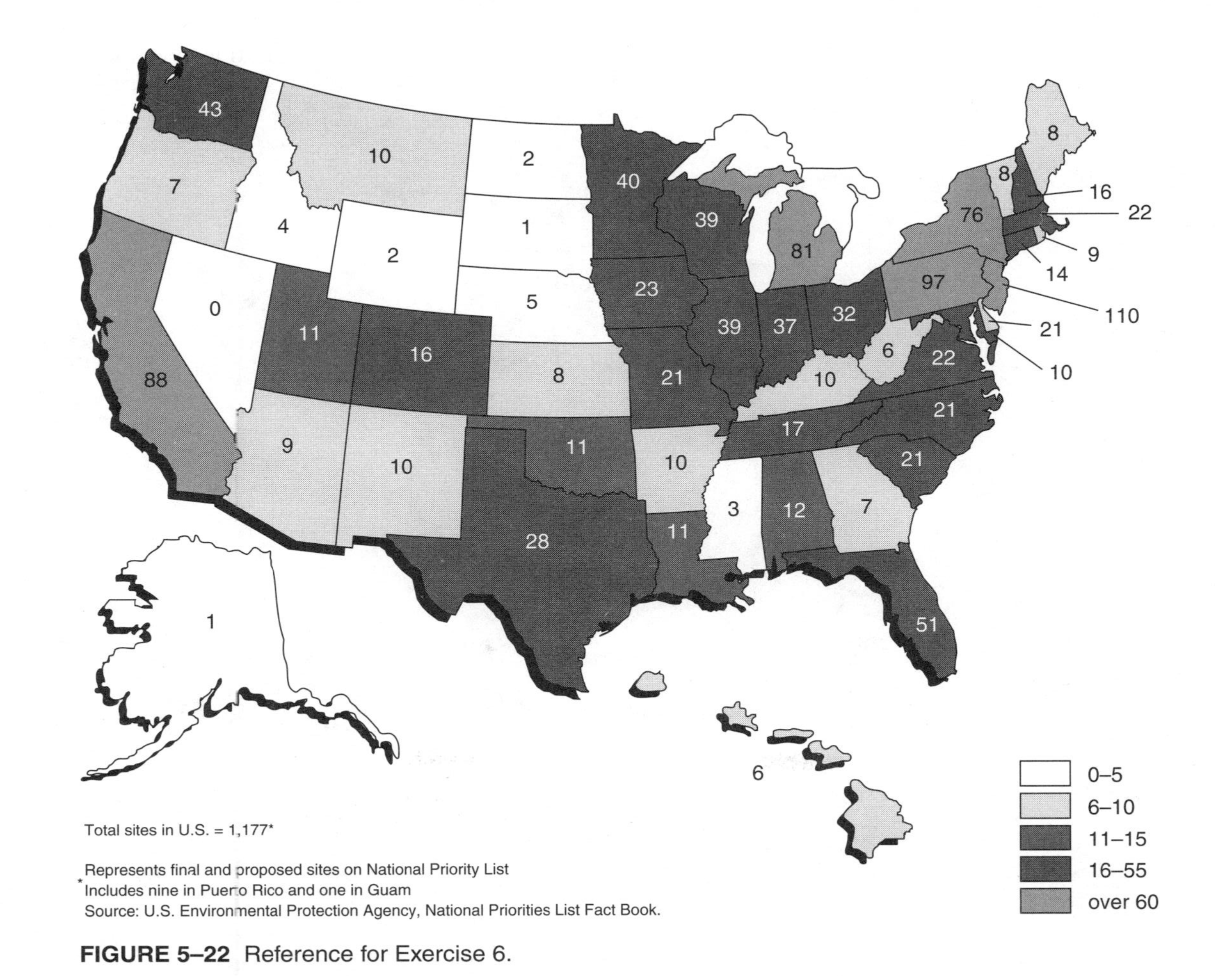

Total sites in U.S. = 1,177*

Represents final and proposed sites on National Priority List
*Includes nine in Puerto Rico and one in Guam
Source: U.S. Environmental Protection Agency, National Priorities List Fact Book.

FIGURE 5–22 Reference for Exercise 6.

7. **Misuse of Graphics**
   Find three deficient graphics in newspapers, magazines, reports, or other documents. Submit copies of the graphics along with an analysis that describes the deficiencies and offers suggestions for improvement.

6 Delivery

You have examined the needs of your audience, researched the topic, organized your material, produced an outline, written the text, and prepared your graphics. You're almost ready for the speech. What remains is your final preparations, when you can select your note format and then practice. This chapter begins with an overview of this final stage. Then it presents guidelines for delivering the speech, handling questions, and reducing speech anxiety.

Some speakers work hard on almost every stage of their speech but then fail to practice enough. Considering the importance of speeches to your personal and professional life, the extra time spent practicing is certainly warranted and usually pays off well. Following are some common situations in which your skill at delivering a speech can lead to success for you or your organization, or both:

- **Getting hired:** A hiring committee may ask you to give a short talk summarizing your education, experience, and career goals.
- **Getting customers:** A potential client may ask you to speak on the highlights of a large proposal you recently submitted.
- **Keeping customers:** A long-term client may ask you to give a speech to the client's employees on technological advances in your products or services.
- **Contributing to your profession:** You may give a speech (sometimes called a "paper") at an annual meeting of your professional association.
- **Contributing to your community:** You may be asked to give a speech on your profession to a community group.

In each case the success of the presentation will require careful attention to the main topics of this chapter: making final preparations, delivering the presentation, and reducing nervousness.

FINAL PREPARATIONS

Two steps precede delivery of the speech and are essential for your success. First, you choose the form in which you write notes to be used during the speech. Second, you must practice, practice, and practice some more. This section offers suggestions on completing both tasks.

OPTIONS FOR NOTES

This book has emphasized that the best speeches usually are "extemporaneous," in which the speaker shows great familiarity with the material and uses notes for reference. Avoid either reading a speech verbatim or giving it totally from memory. In the first case you are ignoring your listeners, and in the second you risk being viewed as an "automaton" who is speaking mindlessly. Even if you know your speech by heart, it's best to look down at your notes occasionally to appear more natural in your delivery.

Besides helping you appear more natural, extemp speaking allows you to make last-minute changes in phrasing and emphasis that might improve delivery. In this way you are not locked into specific phrasing that is memorized or written out word for word. Following are some methods for recording notes to be used during your speech.

Method 1: Marginal Notes in Text

Some speakers gain confidence by having precise, typed text with them during the speech, even though they will not be reading it. In the margins of the paper they have written the notes they will use while speaking. Thus, the text is available to them until the last minute as a reference for matters of wording. However, this option allows little room for notes in the margins. Plus the presence of the full text may tempt you to resort to reading verbatim,in spite of your best intentions.

Method 2: Outline on Paper

Some speakers prefer an outline because it allows them to see the complete expanse of the speech material on one or two sheets of paper. The format of an outline—with its fragmented points arranged in an indented hierarchy—gives the eye an easy reference point during the speech. The outline also assists you in avoiding the distraction of flipping notecards (see the next method). However, reliance on sheets of paper makes it more difficult to move away from a lectern unless you are comfortable using one hand to hold a clipboard or other device for outline sheets.

Method 3: Notecards

The most traditional vehicles for speech notes are 3×5 inch or 4×6 inch notecards. Their main advantages are (1) readability, in that just one or two points are on each card, (2) convenience, in that they are easy to carry and store in a pocket or purse, (3) ease of revision, in that you can add, delete, or change the order of cards, and (4) mobility, in that you can easily hold them in your hand as you move beyond the lectern. Yet there are also some disadvantages. For example, an inexperienced speaker may cause a distraction by flipping through cards noisily. And, as already noted with respect to outlines, cards don't provide you with a full view of the entire speech. If you use cards, make certain to number them for easy placement in correct sequence.

Method 4: Notes on overhead transparencies or PowerPoint

If you select this increasingly popular option, you place a running outline of your presentation on a series of transparencies or PowerPoint slides. The graphics keep listeners attentive and also free you from the need to carry cards or hard-copy outlines. Of course, any dependence on technology can be risky. In case the equipment malfunctions, always have a hard copy of the note slides with you.

Options for Practice

Many speakers prepare a well-organized speech but then fail to add the essential ingredient: practice. Constant practice distinguishes superior presentations from mediocre ones. It also helps to eliminate the nervousness that most speakers feel at one time or another. In practicing your presentation, make use of four main techniques. They are listed here, from least effective to most effective:

Practice Option 1: Practice Before a Mirror

This old-fashioned approach allows you to hear and see yourself in action. The drawback, of course, is that it is difficult to evaluate your own performance while you are speaking. Nevertheless, such run-throughs definitely make you more comfortable with the material.

Practice Option 2: Use an Audiotape

Most presenters have access to a tape player, so this approach is quite practical. The portability of the machines allows you to practice almost anywhere. Although taping a presentation will not improve gestures, it will help you discover and eliminate verbal distractions such as filler words (uhhhh, um, ya know).

Practice Option 3: Use a Live Audience

Groups of your colleagues, friends, or family—simulating a real audience—can provide the kinds of responses that approximate those of a real audience. In setting up this type of practice session, however, be certain that observers understand the criteria for a good presentation and are prepared to give an honest, forthright critique.

Practice Option 4: Use a Videotape

This technique allows you to see and hear yourself as others do. Your careful review of the tape, particularly when done with another qualified observer, can help you identify and eliminate problems with posture, eye contact, vocal patterns, and gestures. At first it can be a chilling experience, but soon you will get over the awkwardness of seeing yourself on tape.

GIVING THE SPEECH

You have prepared the speech and practiced it often. If you have a tendency to get nervous before such occasions, you are employing strategies for reducing anxiety (see the next section). Now it's "showtime." That term in fact does apply because each presentation is actually a heightened form of normal speech. Although you want to appear natural to the audience, you do not want to display a casual manner. First and foremost, you are obligated to keep people interested in the speech and must use techniques that are not evident in normal, informal talk. Toward this end, this section includes guidelines for delivering the speech and handling questions from the audience.

GUIDELINES FOR DELIVERY

Follow the guidelines in this section to make each speech "showtime," without making it "showy." With practice, you will discover and refine the delivery techniques that work best for you.

Presentation Guideline 1: Speak Vigorously and Deliberately

"Vigorously" means with enthusiasm; "deliberately" means with care, attention, and appropriate emphasis on words and phrases. The importance of this guideline becomes clear when you think back to how you felt during the last speech you heard. At the very least, you expected the speaker to show interest in the subject and to demonstrate enthusiasm. Good information is not enough. You need to arouse the interest of the listeners.

You may wonder, "How much enthusiasm is enough?" The best way to answer this question is to hear or preferably watch yourself on tape. Your delivery should incorporate just enough enthusiasm so that it sounds and looks a bit unnatural to you. Few if any listeners ever complain about a speech being too enthusiastic or a speaker being too energetic. But many, many people complain about dull speakers who fail to show that they themselves are excited about the topic. Remember—every presentation is, in a sense, "showtime."

Presentation Guideline 2: Avoid Filler Words

Avoiding filler words presents a tremendous challenge to many speakers. When they think about what comes next or encounter a break in the speech, they may tend to fill the gap with filler words and phrases such as these:

uhhhhh . . .

ya know . . .

okay . . .

well . . . uh . . .

like . . .

I mean . . .

umm . . .

These gap-fillers are a bit like spelling errors in written work. Once your listeners find a few, they start looking for more and are derailed from your presentation. To eliminate such distractions, follow these three steps:

- **Step 1:** Use pauses to your advantage. Short gaps or pauses inform the listener that you are shifting from one point to another. In signaling a transition, a pause serves to draw attention to the point you make right after the pause. Note how listeners look at you when you pause. Do not fill these strategic pauses with filler words.
- **Step 2:** Practice with tape. Tape is brutally honest: When you play it back, you will become instantly aware of fillers that occur more than once or twice. Keep a tally sheet of the fillers you use and their frequency. Your goal will be to reduce this frequency with every practice session.
- **Step 3:** Ask for help from others. After working with tape machines in Step 2, give your speech to an individual who has been instructed to stop you after each filler. This technique gives immediate reinforcement.

Presentation Guideline 3: Use Rhetorical Questions

Enthusiasm, of course, is the best delivery technique for capturing the attention of the audience. Another technique is the use of rhetorical questions at pivotal points in your presentation.

Rhetorical questions are those you ask to get listeners thinking about a topic, not those you would expect them to answer out loud. They prod listeners to think about your point and set up an expectation that important information will follow. Also, they break the monotony of standard declarative sentence patterns. For example, here is a rhetorical question used by a computer salesperson in proposing a purchase by a firm:

> I've discussed the three main advantages that a centralized word processing center would provide your office staff. But you may be wondering, "Is this an approach we can afford at this point in the company's growth?"

Then the speaker would follow the question with remarks supporting the position that the system is affordable.

"What if" scenarios provide another way to introduce rhetorical questions. They gain listeners' attention by having them envision a situation that might occur. For example, a safety engineer could use the following rhetorical question in proposing asbestos-removal services to a regional bank:

"What if you repossessed a building that contained dangerous levels of asbestos? Do you think your bank would then be liable for removing all of the asbestos?"

Rhetorical questions do not come naturally. You must make a conscious effort to insert them at points when it is most important to gain or redirect the attention of the audience. Three particularly effective uses follow:

- **As a grabber at the beginning of a speech:** "Have you ever wondered how you might improve the productivity of your word processing staff?"
- **As a transition between major points:** "We've seen that centralized word processing can improve the speed of report production, but will it require any additions to your staff?"
- **As an attention-getter right before your conclusion:** "Now that we've examined the features of centralized word processing, what's the next step you should take?"

Presentation Guideline 4: Maintain Eye Contact

Your main goal—always—is to keep listeners interested in what you are saying. This goal requires that you maintain control, using whatever techniques possible to direct the attention of the audience. Frequent eye contact is one good strategy.

The simple truth is that listeners pay closer attention to what you are saying when you look at them. Think how you react when a speaker makes constant eye contact with you. If you are like most people, you feel as if the speaker is speaking to you personally—even if there are 100 people in the audience. Also, you tend to feel more obligated to listen when you know that the speaker's eyes will be meeting yours throughout the presentation. Here are some ways you can make eye contact a natural part of your own strategy for effective oral presentations:

- **With audiences of about thirty or less:** Make regular eye contact with everyone in the room. Be particularly careful not to ignore members of the audience who are seated to your far right and far left (see Figure 6–1). Many speakers tend to focus on the listeners within Section B. Instead, make wide sweeps so that listeners in Sections A and C get equal attention.
- **With large audiences:** There may be too many people or a room too large for you to make individual eye contact with all listeners. In this case, focus on just a few people in all three sections of the audience noted in Figure 6–1. This approach gives the appearance that you are making eye contact with the entire audience.
- **With any size audience:** Occasionally look away from the audience—either to your notes or toward a part of the room where there are no faces

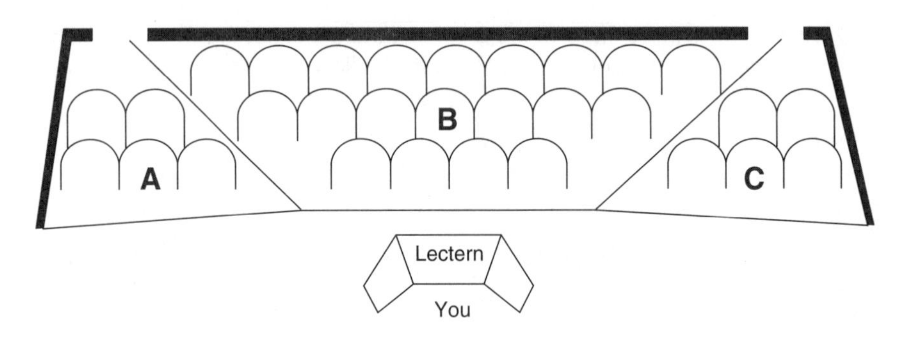

FIGURE 6–1
Audience sections.

looking back. In this way, you avoid the appearance of staring too intensely at your audience. Also, these breaks give you the chance to collect your thoughts or check your notes.

Presentation Guideline 5: Use Appropriate Gestures and Posture

Speaking is only one part of giving a speech; another is adopting appropriate posture and using gestures that will reinforce what you are saying. Good speakers are much more than "talking heads" before a lectern. Instead, they do the following:

1. Use their hands and fingers to emphasize major points
2. Stand straight, without leaning on or gripping the lectern
3. Step out from behind the lectern on occasion, to decrease the distance between them and the audience
4. Point toward visuals on screens or charts, without losing eye contact with the audience

The audience will judge you by what you say and what they see, a fact that again makes videotaping a crucial part of your preparation. By working on this aspect of your speech delivery, you naturally will avoid problems like keeping your hands constantly in your pockets, rustling pocket change (remove it and your keys from pockets beforehand), tapping a pencil, scratching nervously, slouching over a lectern, and shifting from foot to foot.

Guidelines for the Question-and-Answer Period

Many presentations close with a period during which you answer questions put forth by the audience. This "unscripted" part of the speech can be most illuminating for listeners, who now have the chance to (1) seek clarification of material you covered, (2) ask for additional information about the subject, or (3) disagree with points you raised. View this session as an opportunity for genuine communication between you and the audience. Following are a few guidelines to assist in that effort.

Q&A Guideline 1: Let the Person Complete the Question

Speakers who are familiar with their material sometimes tend to anticipate questions and cut people off before they have finished asking the question. Besides being discourteous behavior, this approach sometimes causes the speaker to make false assumptions about the question and, thus, to miss target on the answer. Let the audience member finish the question. In those rare instances when a person talks too long without zeroing in on a question, you can offer a mild corrective such as, "I'm not sure I understand where you're going with that question . . . " or the like.

Q&A Guideline 2: Be Sure You Understand the Question

Some questions are poorly phrased. If you're not sure what has been asked, request that the person rephrase the question. Another approach is to rephrase it yourself before you answer it. For example, you might respond by saying, "If I understand you correctly, you want to know if the insurance premiums are likely to go up more in the next five years than they have in the last five. Do I have that right?"

Q&A Guideline 3: Repeat the Question if the Audience Didn't Hear It

Listeners get frustrated when speakers begin answering questions that were not heard by the entire audience. If you have any doubt that the question was heard by everyone in the room, repeat it. Of course, you can simply ask if everyone heard the question, but be aware that some reticent listeners won't tell you even if they did not. It's best to repeat the question if there is any doubt.

Q&A Guideline 4: Pause for a Moment Before You Begin Your Answer

This technique helps you gather your thoughts before you launch into an answer. The extra few moments are crucial for making sure you have an answer and creating a quick outline in your head. The pause also gives listeners the impression that you have carefully considered the question, as of course you have.

Q&A Guideline 5: Admit It When You Don't Know an Answer

Naturally, most of us don't like to admit when we lack knowledge about a topic about which we have been asked to speak. But invariably you will encounter questions that you simply are not able to answer. In such cases it is always best to say right away that you do not have the information needed. Then tell questioners that you will locate the answers later and get back to them. When you say you will follow up, of course, always do so.

Q&A Guideline 6: Avoid Being Defensive

Sometimes you'll face questions or comments to which your natural inclination might be to respond defensively or to debate the point. If debating is the purpose of the event, then proceed. But most speeches are not debates, and a quarrelsome exchange only detracts from the impact of the speech. Do your best not to respond in kind to those who are trying to engage you in argument. Instead, you can (1) acknowledge the differences of opinion openly, (2) agree to disagree on the point, or (3) offer to discuss the point after the speech.

Q&A Guideline 7: Always End On Time

Simply because there are still questions being asked does not mean you should go beyond the time allotted. Follow the time limit of the Q&A period just as rigorously as you stay within the time limits for your speech. If there are still questions, you might want to suggest that the individuals contact you by e-mail or phone for further information.

DEALING WITH NERVOUSNESS

At some time in their lives, most speakers will feel nervous before a presentation. It's perfectly normal. Indeed, the best speakers are often those who learn to use nervousness to their advantage, rather than letting it become an obstacle. This section includes background information about the problem of speech anxiety as well as concrete suggestions to help you stay calm. Remember as you read, however, that you cannot eliminate all anxiety, nor should you try to do so. Some level of moderate anxiety—as opposed to "high anxiety"—keeps your speech engaging and your tone energetic.

REASONS FOR SPEAKING ANXIETY

For many people, an instinctive "fight or flight" response kicks in when they speak. It may even engage when they first learn they will be giving a speech, sometimes building to a point where the anxiety is outrageously out of proportion to the importance of the speech. To deny this reality would be to ignore an important issue in public speaking. Surveys have shown that people place public speaking near the top of their list of fears, even above sickness and death!

Despite this prevailing sense of concern about public speaking, most of us feel comfortable conversing in smaller groups, such as when talking with friends or colleagues in impromptu or scheduled sessions. We're used to situations that involve the informal exchange of ideas. Formal presentations, however, place us in a more structured, more awkward, and more stressful environment. Although we may know the audience is interested in what we have to say, the formal context triggers nervousness that is sometimes difficult to control.

As already noted, your ultimate goal is to reduce nervousness to the point where it serves your purpose—that is, where it helps you create an animated, enthusiastic presentation. Just as veteran actors use some degree of nervousness to improve their performance, speakers aiming for excellence can benefit from the same effect. The guidelines that follow will help you address the problem.

Zen and the Art of Reducing Anxiety

As the cliché goes, don't try to eliminate "butterflies" before a presentation; just get them to fly in formation. It's best to acknowledge that a certain level of nervousness will remain even after you've applied the guidelines in this chapter. Then you can go about the business of getting it to work for you, not against you.

The title of this section—a play on the title of Robert Pirsig's excellent 1971 book, *Zen and the Art of Motorcycle Maintenance*—does not suggest this chapter contains a serious discussion of Zen Buddhism; it does not. Yet there are features of Zen that do apply to public speaking. You may know that Zen involves a kind of quiet meditation that aims to reduce sensory overload in our lives. Just sitting and focusing on a word (mantra) or a physiological activity (like breathing) helps to calm down a racing body. In applying the term "Zen" somewhat loosely, therefore, the guidelines will help you take control of any speaking situation and to develop self-confidence. Make the following guidelines part of your speech ritual.

No-Nerves Guideline 1: Prepare Well

The most obvious suggestion for eliminating nervousness is also the most crucial one. Work hard to prepare your speech so that your command of the material will help to conquer any queasiness you feel. Besides making you feel more confident, knowledge of the material helps listeners overlook any nervousness you think may be evident.

Although this guideline is one over which speakers have the most control, many disregard it for two quite different reasons. Some believe they know the material so well that they don't need to practice the speech. After all, haven't they been working in this field all of their college years or throughout their careers? These speakers often have a problem different from, but probably worse than, excessive nervousness—a rambling speech.

A second group of speakers fails to prepare for a reason more related to this topic. They are so nervous about the speech that they reach a point of resignation. They surmise that nothing can help them at this point, so they develop a sense of doom that aggravates any anxiety that may develop.

You must resist any temptation to "wing it" in a speech and thus reject the faulty thinking just noted. Instead, *overprepare* to the degree that you know your material inside and out. Your high level of knowledge will add to a level of confidence that combats nervousness.

No-Nerves Guideline 2: Prepare Yourself Physically

In medieval Japan, many samurai warriors adopted the Zen Buddhist philosophy because it reinforced the importance of mental and physical preparation before battle. Although it may seem to be a "stretch" to compare giving a speech to preparing for battle, some legitimate parallels do exist. As noted earlier, the "fight or flight" instinct often kicks in when a speaker experiences a high degree of nervousness caused by rushing adrenaline. Physical activity can help to relax this reaction.

You may know of athletes who follow a certain routine before a performance, such as eating a particular meal they think best prepares them for competition or warming up with a particular sequence of exercises. Some successful speakers view the physical preparation before a speech in the same way. Here is a compilation of some common precautions:

- **Avoid ingesting caffeine or alcohol for several hours before you speak.** Both of these drugs can alter your response even in small amounts—either by slowing down your reactions in the case of alcohol or speeding them up in the case of caffeine. Just as important, they may add to your anxiety by causing you to question whether, in fact, they might be influencing you to some degree. Again, avoid any substance, activity, or thought that may reduce self-confidence.
- **Eat a light, well-balanced meal within a few hours of speaking.** Most people prefer a balanced meal an hour or two before a speech to keep the body functioning smoothly. However, don't overdo it by eating heavily—particularly if the meal comes right before your performance as a featured speaker.
- **Perform deep-breathing exercises before you speak.** This technique mirrors the ancient Zen practice of focusing on the act of breathing. Start by becoming conscious of your breathing, and then work at inhaling and exhaling slowly. In this way you train your body to slow down to a pace you can control. Exerting some control over an activity considered involuntary takes your mind off the speech. It brings your body into a state of rest not unlike that of Zen monks who meditated while sitting cross-legged on straw mats. We can learn from the ancient traditions.

- **Exercise normally the day of the presentation.** A good walk, for example, may help to invigorate you and use up the adrenaline that leads to nervousness before a presentation. Or you may prefer more vigorous exercise such as swimming, biking, or jogging. However, don't wear yourself out by exercising in what would be an unusually excessive fashion for you.
- **Drink several glasses of water in the hours before you speak.** When you properly hydrate your body, your vocal cords are most likely to operate without strain during a presentation. Of course, for obvious reasons, don't take this suggestion to the extreme by drinking excessively before speaking.

No-Nerves Guideline 3: Picture Yourself Giving a Great Presentation

Some view this guideline as a variation of an athlete's efforts to put on a "game face" before competing. Others may see it as analogous to the technique of "visualizing" success. The point is to flood your psyche with images of giving a successful presentation before a group of listeners who are being informed, persuaded, or entertained by your words. Whatever other value this technique may have, it certainly does take your mind off anxiety by keeping you preoccupied with the effort to "visualize."

As part of the process of picturing yourself giving a successful speech, take yourself through the following sequence of actions that make up the presentation:

- Arrive at the room with prepared materials in hand.
- Check the visuals to see that equipment is functioning properly.
- Exchange pleasantries with other presenters or members of the audience.
- Rest comfortably in your chair while you are introduced.
- Stand up and stay silent for a few seconds while you prepare to start.
- Begin your speech with a smile and an attention-getting introduction.
- Maintain eye contact and an appropriate pace throughout the speech.
- Conclude with an effective wrap-up.
- Respond with clarity and thoroughness to all questions.
- Re-take your seat and know that you have "delivered the goods."

The visualization technique sometimes is called "imaging." Whatever label you give it, however, the point is that it works by programming success into your thinking before you speak. The experience of many who have used this strategy suggests that it helps control negative thoughts that pass through the minds of even the best speakers. As already noted, if nothing else, it occupies your mind with a sequential process at a time that you might otherwise be allowing feelings of nervousness to dominate your mind.

No-Nerves Guideline 4: Arrange the Room the Way You Want

Prescribed rituals were an important part of Zen traditions in medieval Japan, from preparing for battle to conducting the tea ceremony. Part of your ritual to prepare for a speech and, thus, to reduce your anxiety should be to assert control over the physical environment in which you speak. Here are some actions you can take:

- Set up chairs in the pattern you prefer.
- Position the lectern to your taste.
- Make certain the lighting is adequate, especially near the lectern.
- Check the position of the equipment so that it's ready for immediate use.
- Ask that the temperature in the room be adjusted for your comfort.

A speech presents you with one of the few communication contexts in which your comfort is more important than anyone else's. Usually, speakers with a moderate level of nervousness prefer a room to be slightly cooler than listeners want it to be. Although you certainly don't want your audience to be shivering in their chairs, remember that a slightly colder room keeps an audience more attentive than a slightly warmer one, which induces drowsiness.

No-Nerves Guideline 5: Have "Crutches" Available for Emergencies

To help them stay hydrated, even the most relaxed speakers often keep a glass of water nearby when they speak. You may have noticed that speakers usually take a drink of water at strategic points, such as while a question is being asked or while the audience is applauding. For speakers who are a bit nervous, having water available is even more important because a dry throat sometimes accompanies nervousness before and during a presentation. Here are a few other "crutches" that may reduce anxiety before a presentation:

- A supply of handouts that would be used if equipment required for your visuals were to malfunction
- Extra marker pens for the overhead projector, smart board, or flipchart—just in case your host has not checked the supplies

The "be prepared" school of public speaking suggests that any action that reduces anxiety and increases preparedness is worth considering.

No-Nerves Guideline 6: Engage in Casual Banter Before the Speech

This suggestion may be the least "Zen-like" on the list in that Zen practitioners are known for their silence. There must be a good reason for substituting words for peaceful thought right before a speech, and occasionally there is. Such casual conversation with the audience often helps reduce the distance you feel between you and the listeners. This distance often presents a psychological barrier to speakers who envision their audience as potential critics or even "the enemy." Almost all listeners empathize with speakers and want them to do well, and you become acquainted with this goodwill when you talk casually with them before the speech. In addition, you get a chance to exercise your voice and calm your nerves. Then it doesn't seem as much of a transition when you begin your formal speech.

No-Nerves Guideline 7: Remember That You Are the Expert

Having completed their Zen preparations with care, Japanese samurai had to be absolutely convinced they were the best warriors on the field. This confidence often made the difference. Similarly, when you deliver a speech in class or at work, you need to remember that you are providing information that your listeners do not have. Reminding yourself that you are the one with expertise on the subject should help to boost your confidence. Your listeners want to hear what you have to say. Tell yourself, "I'm the expert here!"

No-Nerves Guideline 8: Don't Admit Nervousness to the Audience

No matter how anxious you feel, resist the temptation to admit it to others. First, you don't want listeners to feel sorry for you—that's not an emotion that leads to a positive view of your speech. Second, you may be surprised to learn that your nervousness is almost never apparent to the audience. Although your heart is pounding, your throat feels dry, and your legs are a bit wobbly, few, if any, listeners will observe these symptoms. Why draw attention to a problem that may be unnoticed? Third, you can best overcome anxiety by pushing through it rather than by drawing extra attention to it.

No-Nerves Guideline 9: Slow Down

Out of nervousness, some speakers tend to speak faster than they did during practice. Because this increased pace often appears obvious to your audience, it can reduce your effectiveness as a speaker. Here are some strategies for preventing it:

- Practice with audiotape or videotape so that you are very familiar with how your voice will sound in a formal presentation.
- Constantly remind yourself to slow down during the speech and to use pauses for rhetorical effect.
- Maintain strong eye contact with your audience—some speakers begin to talk too fast when they reduce eye contact with the audience and focus on notes.
- Write reminders on the copy of your speech such as "Slow down here!" or "Whoa!"
- Time yourself in practice, and then note in the margin of the speech text about where you should be at certain intervals (of course, any looking toward your watch during the speech itself should be subtle).

No-Nerves Guideline 10: Seek Assistance

Samurai in medieval Japan knew the value of trusted colleagues and often practiced battle techniques with them. Just as important to them as meditation was the notion that they were forever bound to their colleagues. Likewise, you do not need to go it alone in your effort to overcome anxiety in speaking. Here are a few ways you can get assistance:

- **Join a speaking organization.** One such organization is Toastmasters International, which has chapters all around the world and encourages new ones to be formed where needed. These clubs meet regularly to give members the opportunity to develop and practice their speaking skills in a supportive environment in which everyone wants to improve. Consult your phone book or check the Internet to contact local Toastmaster chapters.
- **Start you own informal speaking organization.** You don't need the umbrella of an international organization to start your own self-help group. Once you have collected a few colleagues who want to improve their speaking skills, set some rules and agree to meet at a regular time without fail—such as every week or two for lunch in the company training room. One or two members can give different types of speeches, followed by constructive discussion by all those present. There is no better way to gain confidence than by speaking more often.
- **Get medical help.** If speaking anxiety presents physical symptoms that concern you, such as a rapid heartbeat, you might want to visit your family doctor for advice on possible medications for such symptoms.

In summary, a moderate case of "nerves" before a speech helps you develop a level of enthusiasm that is welcomed by listeners. However, when you have anxiety well beyond what you think is useful to your purpose, employ the techniques in this chapter to get your butterflies to "fly in formation."

CHAPTER SUMMARY

This chapter covers the speech-related tasks that occur at the end of the process but are essential to the success of your effort. First, it addresses the choices you have for recording notes to be used in an extemporaneous speech. Also included are descriptions of the main practice methods—using a mirror, an audiotape, a live audience, or a videotape. Then the chapter moves to specific guidelines for delivery that concern criteria such as tone, diction, eye contact, and gestures. Brief guidelines for handling questions are given. The chapter concludes with a discussion of speech anxiety and a list of suggestions for reducing its effect on your presentations.

EXERCISES

Your instructor will indicate the time limits for all of the following assignments. For each one you could use the guidelines in this chapter, especially those concerning speech notes and delivery.

1. **Speech on Your Academic Major**
   Prepare and give a presentation in which you discuss (a) your major field, (b) reasons for your interest in this major, and (c) specific career paths you may pursue related to the major. Assume your audience is a group of students with undecided majors, who may want to select your major. Use at least three graphics.
2. **Speech Based on a Written Report**
   Using a written assignment you have prepared for a recent course, prepare a speech based on the report. Assume that your main goal is to present the audience with highlights of the written report, which you should assume listeners have not yet read. Use at least three graphics.
3. **Speech Based on a Proposal**
   Prepare a presentation based on a written proposal you have written for a college course or for a job. Assume your listeners are in the position of accepting or rejecting your proposal but that they have not yet read the written document.
4. **Speech Based on a Formal Report**
   Prepare a group or panel presentation using the size groups indicated by your instructor. The speech may be related to a collaborative writing assignment or done as a separate project. In planning your panel, be sure that group members move smoothly from one speech to the next, creating a unified effect.
5. **Group Presentation: Internet Search**
   Prepare a group presentation that results from work your group does on the Internet. Retrieve information about one or more businesses or careers in a particular country. Once you have split up the group's initial tasks, conduct some of your business by e-mail. Then present the results of your investigation in a panel presentation to the class. For example, your topic could be the computer software industry in England, the tourist industry in Costa Rica, or the textile industry in Malaysia.

6. **Speech Based on Exercises in Chapter 2**
   Prepare and deliver a speech based on Exercise 3, Exercise 4, or Exercise 5 in chapter 2 (Research). These exercises ask you to select or develop topics related to your personal experience or to your research.
7. **Speech Based on Exercises in Chapter 3**
   Prepare and deliver a speech based on the results of Exercise 2, Exercise 3, or Exercise 4 in chapter 3 (Organization). These exercises ask you to develop outlines for informative, persuasive, and occasional speeches.

Appendix: Sample Speech

Although no written speech can help you develop the delivery techniques discussed in chapter 6, the speech text *can* provide a model for organizing information and shaping prose. It is for such a purpose that this appendix includes a short speech entitled "Asbestos: Why You Need to Know."

Kim Mason, an industrial hygienist with a consulting firm in Atlanta, has been asked to give a short, after-dinner presentation to a group of building owners on the problem of asbestos contamination. The members of Kim's audience have an obvious interest in the problem—they own buildings that may be at risk. Yet most of them know little about asbestos except that it's a health issue they must consider when they renovate their buildings. Kim's job is to heighten their awareness, without appearing to be pushing the services of her own employer (which does asbestos-abatement work). Kim believes, rightly, that she has an ethical obligation to avoid overt promotion of her own firm's services when she has been invited to provide technical information. Here is her speech.

ASBESTOS: WHY YOU NEED TO KNOW

Kim Mason, C.I.E.

[1] Good evening. My name is Kim Mason, and I work for the asbestos-abatement division of Runzach, Inc., in Atlanta. I've been asked to give a short presentation on the problem of asbestos, and then to respond to your questions about the importance of removing it from buildings.

PIP formula begins—Clearly states purpose.

[2] My interest in the topic began ten years ago when I completed a college research paper on the long-term medical problems associated with asbestos contamination.

Creates interest by noting her background in field.

Shortly thereafter I began training and coursework that eventually led to my becoming a Certified Industrial Hygienist. Now a good deal of time is spent investigating potentially hazardous sites that might contain asbestos and then recommending remediation.

Forecasts main points of the speech.

[3] The topic of asbestos is scientifically interesting and technically complex. To give it the most relevance to your business—owning and managing buildings—I thought it would be most useful to focus on three main reasons why you, as building owners, should be concerned about the asbestos problem: (Transparency 1)

Uses overhead to present three main points. Focuses on limited number.

1. To prevent future health problems of your tenants
2. To satisfy regulatory requirements of the government
3. To give yourself peace of mind for the future

Again, my comments will provide just an overview, serving as a basis for the question session that follows in a few minutes. (Transparency 2)

Uses Q&A pattern to structure body of the speech. Starts with the most important point, for emphasis.

[4] Question: What is the most important reason you need to be concerned about asbestos? Answer: The long-term health of the tenants, workers, and other people in buildings that contain asbestos. Research has clearly linked asbestos with a variety of diseases, including lung cancer, colon cancer, and asbestosis (a debilitating lung disease). Although this connection was first documented in the 1920s, it has only been taken seriously in the last few decades. Unfortunately, by that time asbestos had already been commonly used in many building materials that are part of many structures today.

Supports point by giving specific examples of asbestos products.

[5] Here are some of the most common building products containing asbestos. (Transparencies 3, 4, and 5) As you can see, asbestos was used in materials as varied as floor tiles, pipe wrap, roof felt, and insulation around heating systems. An abundant and naturally occurring mineral, asbestos was fashioned into construction materials through processes such as packing, weaving, and spraying. Its property of heat resistance, as well as its availability, was the main reason for such widespread use.

Uses graphics to make transition to next main point.

[6] While still embedded in material, asbestos causes no real problems. However, when it deteriorates or is damaged, fibers may become airborne. In this state, they can enter the lungs and cause the health problems mentioned a minute ago. This risk prompted the Environmental Protection

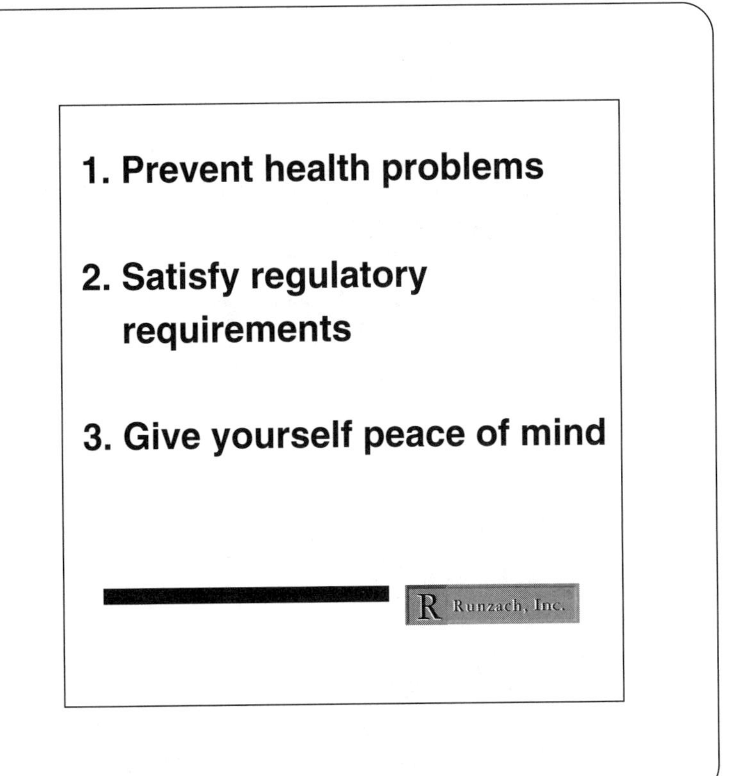

Transparency 1 for sample presentation

Transparency 2 for sample presentation

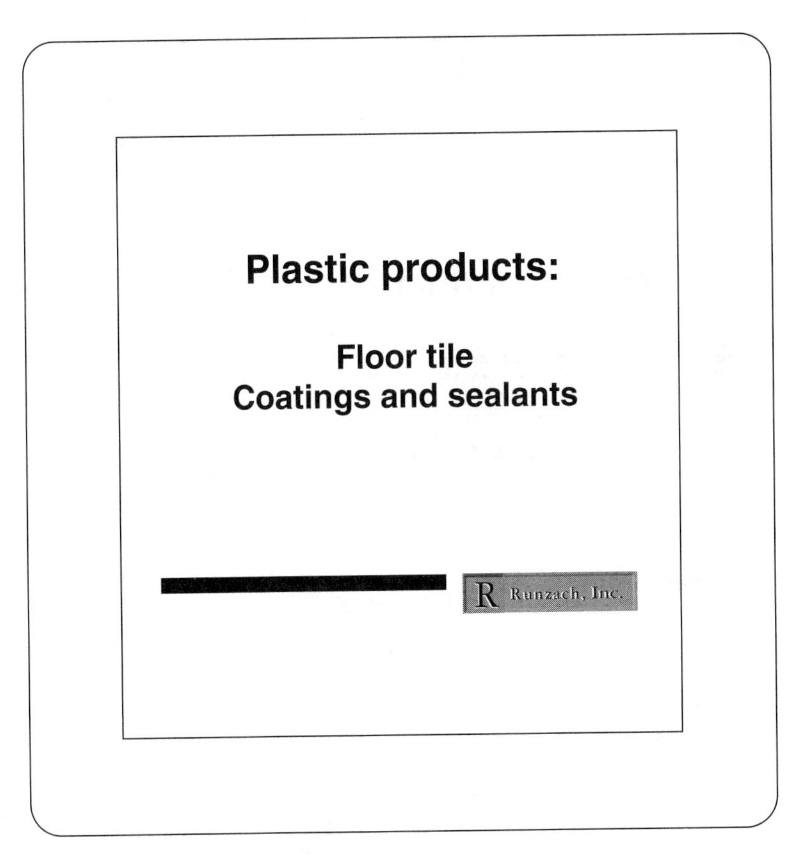

Transparency 3–5 for sample presentation

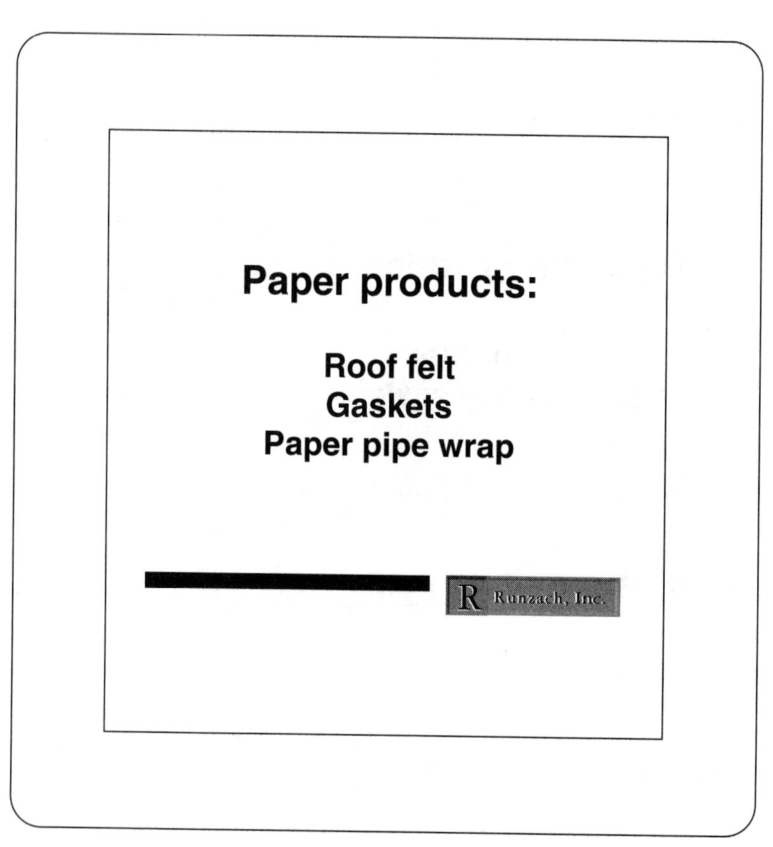

Transparency 3–5 for sample presentation

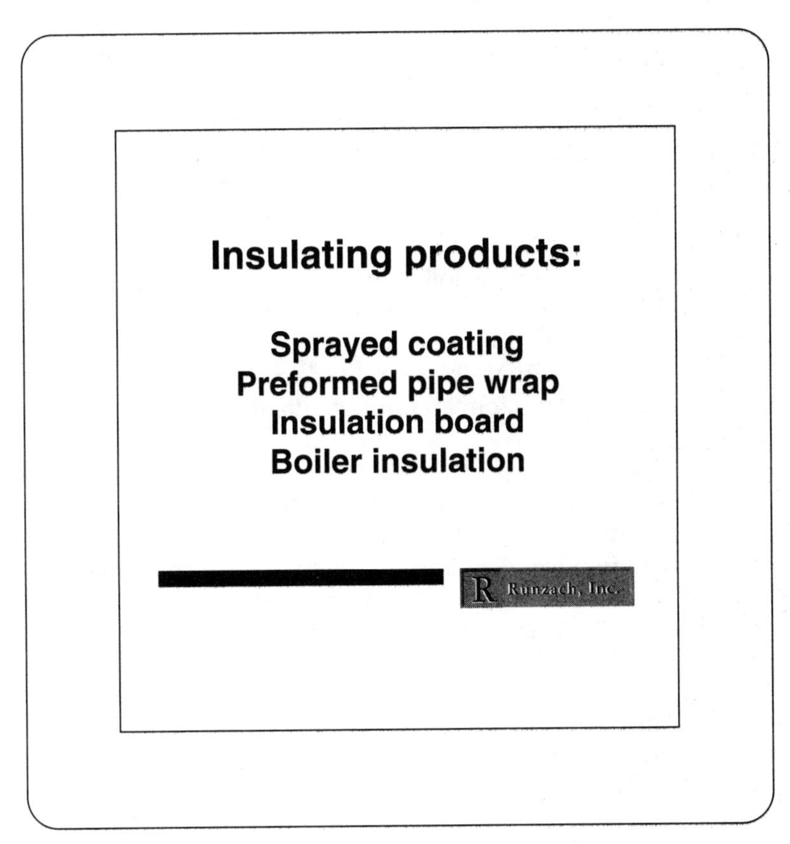

Transparency 3–5 for sample presentation

Agency in the mid-1970s to ban the use of certain asbestos products in most new construction. But today the decay and renovation of many asbestos-containing building materials may put many of our citizens at risk for years to come. (Transparency 6)

Again uses rhetorical question to introduce a main point. Defines abbreviations. Provides specific examples that relate to listeners' experience.

[7] After your concern for occupants' health, what's the next best reason to learn more about asbestos? It's the law. Both OSHA (Occupational Safety and Health Administration) and the Georgia DNR (Department of Natural Resources) require that you follow certain procedures when structures you own could endanger tenants and asbestos-removal workers with contamination. For example, when a structure undergoes renovation that will involve any ACM (which stands for asbestos-containing material), the ACM must be removed by following approved engineering procedures. Also, the contaminated refuse must be disposed of in approved landfills. Considering the well-documented potential for health problems related to airborne asbestos, this legislative focus on asbestos contamination makes good sense.

[8] By the way, OSHA and DNR regulations require removal of asbestos by licensed contractors. But these contractors will assume liability only for what they have been told to remove. They may or may not have credentials and training in health and safety. Therefore, building owners should hire a firm with a professional who will (1) survey the building and present a professional report on the degree of asbestos contamination and (2) monitor the work of the contractor in removing the asbestos. By taking this approach, you as an owner stand a good chance of eliminating all problems with your asbestos.

Engages interest by showing why they *must* be concerned—for their own self-interest.

[9] Yes, it is your asbestos. As owner of a building, you also legally own the asbestos associated with that building—forever. For example, if a tenant claims to have been exposed to asbestos because of your abatement activity and then brings a lawsuit, you must have documentation showing that you contracted to have the work performed in a "state-of-the-art" manner. If, as recommended, you have hired a qualified monitoring firm and a reputable contracting firm, liability will be focused on the contractor and the monitoring firm—not on you. (Transparency 7)

Transparency 6 for sample presentation

Transparency 7 for sample presentation

[10] What, then, is the last reason for concerning yourself with any potential asbestos problem? The answer is—peace of mind. If you examine and then effectively deal with any asbestos contamination that exists in your buildings, you will sleep better at night. For one thing, you will have done your level best to preserve the health of your tenants. For another, as previously noted, you will have shifted any potential liability from you to the professionals you hired to solve the problem—assuming you hired professionals. Your monitoring firm will have continuously documented the contractor's operations and will have provided you with reports to keep in your files, in the event of later questions by lawyers or regulatory agencies.

Shows how practical and ethical reasons merge.

[11] In just these few minutes, I have given only highlights about the three main reasons you need to know more about asbestos (Transparency 1)—to prevent health problems of your tenants, to satisfy the regulators at all levels, and to give yourself peace of mind. It poses a considerable challenge for all of us. Yet the current diagnostic and cleanup methods are sophisticated enough to suggest that the problem, over time, can be solved. If you're wondering what to do next, my suggestion is that gathering more information might be a good start. On the back table I've placed a list of book references and Web sites you can consult. Feel free to take a copy as you leave this evening. Now, I would be glad to answer your questions.

Restates three main points of speech. Indicates what should occur next.

Index